MICROBIC DISSOCIATION II
TWORT-D'HERELLE

CRITICAL REVIEW AND A NEW (HOMOGAMIC) CONCEPTION OF BACTERIOPHAGE ACTION

PHILIP HADLEY

From the Hygenic Laboratory
University of Michigan, Ann Arbor
1928*

I0478324

FOREWORD
S. H. SHAKMAN
INSTITUTE OF SCIENCE
Santa Monica, CA USA
InstituteOfScience.com
I-o-S.org
2013

*Reprinted from *The Journal of Infectious Diseases*, Vol. 42, No. 4 (1928).
Pages 1-172 herein correspond to pages 263-434, respectively, of the original.

Published in the USA
by the Institute Of Science, Santa Monica, CA

FOREWORD
S. H. Shakman

BACTERIOPHAGE – AUTO-ANTIBODY?

Nearly a century has passed since publication of the original papers by Twort in 1915 and d'Herelle in 1917 on the submicroscopic entities known as bacteriophages and their behavior, commonly referred to as the Twort-d'Herelle Phenomenon.

> * Twort, F. W.: An investigation into the nature of the ultramicroscopic viruses, Lancet, 1915, 2, p. 124.
> * d'Herelle, F.: Sur un microbes invisible antagoniste des bacilles dysenterique, Compt. rend. Acad. d. Sc, 1917, 165, p. 373.

But surprisingly, notwithstanding the passage of time and all the sophisticated developments of the ten decades following, the Twort-d'Herelle phenomenon is receiving renewed, increased and serious attention today. The great promise of the antibiotic age, ushered in with the discovery of penicillin and its kin, has been not quite been broken but certainly has been moderated, seemingly through overuse and/or the emergence of antibiotic-resistant strains of pathogens. And the currently-dominant medical drug craze, which the antibiotic era arguably has to some extent spawned, has seemingly turned a frightening portion of the medical profession into a gang of profiteering drug middlemen with a general blatant disregard for history and basic physiology; perhaps its greatest success has been to enable prescription drugs to surpass the automobile as the number one cause of unintentional fatalities in the U.S.

Thus, the medical student continues to be indoctrinated from the outset that the accepted response to harmful effects of one drug is to prescribe a second one that addresses those harmful effects, and so on. And the successes of medical giants such as E.C. Rosenow in extending the venerable principles of Henle etal, through replication of disease conditions in all manners of laboratory animals, and development of specific therapies therefrom, has been replaced by an endless parade of human guinea pig "clinical trials", with a disgusting disrespect for the sanctity of human life and health. Why in the world would any presumably healthy individual (in his or her "right mind") consider enrolling in one of these "clinical trials" for the alleged "benefit of the advancement of science"? Call it what it is – cynical exploitation of the financial underclass by "clinical trial" charlatans for the benefit of the bottom line of their corporations and medical frontmen. Or so it seems, but I digress, and beg forgiveness for this rant (or lapse into reality).

In any case, there is certainly room for a more organic, biologically sound approach. Thus, recent times have seen an upsurge of interest in the Twort-d'Herelle phenomenon; its alternative perspective on combating infections seems to some to be the wave of the future.

> - - - William C. Summers, Félix d'Herelle and the Origins of Molecular Biology. 230 pp. New Haven, Conn., Yale University Press, 1999. (Book Review: N Engl J Med, February 24, 2000, 342:595
> - - - Anna Kuchment, The Forgotten Cure: The Past and Future of Phage Therapy, Copernicus Books, 2011
> - - - Borrell, Brendan, "Phage Factor", Scientific American, Aug 2012, Vol. 306 Issue 8, p80-83

In its simplest characterization, Hadley views the bacteriophage not as an entity external to the bacteria it "attacks" if you will, but rather as an intrinsic component of that bacteria which is somehow turned against its progenitor; in contrast to d'Herelle, to whom it was viewed as an infecting virus. In that the d'Herelle view continues to dominate the discussion in the early 21st Century, the contrary position of Hadley, if substantiated, is of central significance – and particularly so given the seeming potential of bacteriophages to take up the therapeutic challenge against which antibiotics seem to be coming up short.

In consideration of the relation between the bacteriophage and microbic dissociation, it seems prudent to briefly review the able introduction to this issue provided by Hadley himself in his earlier work of 1927, Microbic Dissociation. At the outset he urges that "the problem of the bacteriophage (transmissible bacterial autolysis)", be kept in mind throughout the reading of his (earlier 1927) work, even though it is not always specifically mentioned in a given section. Therefore, the reader is directed to the first volume for a full treatment/review of the subject of bacteriophage relative to the primary thread of thought therein (microbic dissociation). For convenience, some selected excerpts from seemingly relevant/example sections (by no means a complete listing) are provided below, i.e., p. 7-8, 24-5, 61-2, 119-20, 163, 214-5, 256.

A brief word on Hadley's writing and wit, notwithstanding the outwardly dry nature of the subject matter: In the first clip below, page 7, we find his overview of the primary difference in views between d'Herelle (re bacteriophage) and Bordet (transmissible bacterial autolysis), their essential difference being involvement of external versus internal action, respectively. While Hadley is certainly more aligned with the latter, what with its easier association with the concept of microbic dissociation, we see later his taking issue with both the d'Herelle general view and Bordet's characterization of the bacteriophage as being able to effect permanent mutations, with the blunt comment that "both of these views are alike in their falsity" (p. 215). But he reserves his harshest critiques for d'Herelle, derisively declaring that "he is apparently oblivious to the existence among bacteria of the phenomenon of microbic dissociation" (p. 214); and pointing out how d'Herelle's special provision for the co-existence of dissociative forms and bacteriophages, which "requires a stretch of imagination for which many bacteriologists may not be qualified", is in essence the strongest argument that might be advanced against d'Herelle's overall position (p. 258).

From Philip Hadley, MICROBIC DISSOCIATION 1927:

-- p. 7 "To all this it may be added that this work has one further purpose which, though not announced in every section, it is hoped will be kept in the mind of the reader throughout. It relates to the problem of the bacteriophage (transmissible bacterial autolysis), about which a further word may be said at this point.* In the field of bacterial autolysis, as we view it today, there are two classes of phenomena demanding special attention. The first of these is microbic dissociation; the second is the classical phenomenon of d'Herelle, termed by him "bacteriophagy," but by Bordet "transmissible bacterial autolysis." Regarding the first, which has received little attention up to the recent time, there is small question regarding its normal, and apparently spontaneous, nature, although the reaction may easily be forced. Regarding the latter, however, there exists a division of opinion regarding the causal agency? one school (that of d'Herelle) maintaining the bacteriophage (Protobios bacteriophagus, syn. Bacteriophagum intestinale d'Herelle, 1918) to be of a living, virus-like nature; the other school (that of Bordet) maintaining the phenomenon to be due to an inherited, vitiated metabolic state of the organism concerned. Whatever may be the relative merits of these respective views, it is perhaps of importance to note that both bacteriophagic action and microbic dissociation (as this subject will be developed in the following pages) have a common meeting ground in a single aspect of physiologic behavior --- namely the instability of the bacterial culture in its reproductive function. In view of this fact it has seemed worth while to conduct a detailed study of phenomena bearing on this subject so far as rendered possible by the literature; and this with a view to ascertaining to what extent, if at all, we can detect any relation in cause, nature or mode of expression, obtaining between these apparently divergent phenomena, proceeding out from a presumably common physiologic state. I believe it is only by a study of such a comprehensive sort that we can derive a sound support for logical and effective methods of investigating the nature and meaning of bacteriophagic action, about which, notwithstanding the most recent and elaborate researches of d'Herelle and his supporters, one must believe the last word has not yet been said.

" If the reader is not already familiar with the phenomenon of the bacteriophage (the d'Herelle phenomenon), much that appertains to this phase of the subject will be to him meaningless. The reading of d'Herelle's most recent (1926) and fully comprehensive work is recommended. [d'Herelle, F.: The bacteriophage and its*

behavior. (Translation) 1926.] For an understanding of the most widely recognized alternative view relating to the theory of the bacteriophage, I would recommend a fairly recent paper by Bordet. [Bordet, J.: Le probleme de l'autolyse microbienne transmissible ou de bacteriophage, Ann. de L'Inst. Pasteur, 1925, 39, p. 717"

-- p. 24-5 "Suicide Cultures : There remains to be mentioned a striking mode of dissociation no doubt observed by many bacteriologists who have had laboratory experience, and this relates to the so-called suicide cultures, which have received mention by Collins but have not obtained a significant place in the literature. Such cultures, after attaining within the first 18 to 20 hours a luxuriant growth, then seem literally to melt away within the space of another 12 to 18 hours until nothing is left but a thin, transparent film covering their former site. To the above a few incidental points may be added. When dissociation occurs in a culture the growth energy always is markedly enhanced; we obtain, as I have pointed out for B. pyocyaneus and some other cultures, a veritable proliferative growth accompanying or preceding the reaction. In connection with lysis accomplished by the bacteriophage in liquid mediums it may also be recalled that Bordet has drawn attention to the necessity of "growth before lysis." D'Herelle also has pointed out this reaction as a characteristic of all cells when attacked by a filtrable virus. [d'Herelle, F.: Immunity in natural infectious disease. (Translation), 1924]"

-- p. 62 "In the case of B. proteus, yellowish colonies possessing unusual characteristics and perhaps representing R type culture, were observed by Braun and Schaeffer. Fejgin also has found five types of B. proteus X19 secondaries to transmissible autolysis which in colony form on agar plates were characterized as being, respectively, white, round, opaque, irregular and chromogenic. Two of the latter type were bright yellow and one was canary yellow, the pigment of all three being soluble in alcohol. In serological tests the antigenic character seems to have remained close to that of the original X19, but in two of the yellow types the agglutinability was much diminished. All were gram-positive and the bright yellow strains were found to contain long filaments along with the rod forms. The growth of the canary yellow strain only was spreading, the organisms small and nonmotile. D'Herelle believed that all of these cultures were "mutations" produced by the bacteriophage. It is clear, however, that the presence of the bacteriophage is not necessary for Mr. Weaver, working in our laboratories, has produced from B. proteus apparently similar yellow strains without the use or presence of the proteus bacteriophage. Whatever the cause of this curious phenomenon may be, there is no doubt that a careful survey of the literature would yield other instances in other bacterial species."

-- p. 119-120 "An apparently similar transformation, but involving B. paratyphosus A, occurring under the influence of the lytic principle, was reported by Bachmann and de la Barrera. In this instance a "mutant" was formed which came to manifest stronger serologic affiliation with B. typhosus than with the original paratyphoid A culture. D'Herelle, [p. 120] in commenting on this case, gives us to understand that the results were caused by the bacteriophage and implies that they can be produced in no other way. Such a view, however, is erroneous as we shall see in section 14 when we consider in greater detail the dissociation-provoking power of the bacteriophage. The relation of the "mutations" described by van Loghem, by Bachmann and de la Barrera and by Baerthlein are apparently all on the same order and furnish support for the conception of bacterial convergence of the R types as first introduced by Schütze and dealt with more conclusively by Goyle, as will be shown on a later page."

-- p 163 "It is of interest, however, that by the use of "bacteriophage suspensions" d'Herelle has been able to immunize with much success against both barbone (Pasteurella bovis) and human plague (B. pestis). It seems clear, from d'Herelle's analysis of the mechanism of the protective reactions here involved, that the bacteriophage is not directly concerned, but that the immunity is due to the generation of protective antibodies possessing opsonic significance. The question arises regarding the extent to which the remarkable results reported may be due to the R type antigen present in the bacteriophage suspensions, i. e., the Pasteurella culture after lysis."

-- p. 214-215 "To the views mentioned above relating to the degree of permanence of the R type cultures arising from active microbic dissociation, a statement should be added regarding the permanence of the presumably analogous R type cultures arising under the influence of the bacteriophage ; for, as we shall see in a later section, the phenomenon of transmissible autolysis, accompanied by its generation of secondary, resistant cultures, in reality may be regarded as a form of dissociation of the original sensitive culture. D'Herelle does not bring out this relation, and for the reason that, throughout all of his publications, he is apparently oblivious to the existence among bacteria of the phenomenon of microbic dissociation which receives no mention, either under this caption or under any other, in any of his works. For him, a resistant culture is always one that has been made so through previous contact with the bacteriophage. He notes, however, that these secondary cultures are often quite different, morphologically, biochemically and serologically, from the original culture. They are all indeed "mutants," immune strains, produced by the bacteriophage! Among them some may revert easily and quickly to the former state while others he regards as permanent in their newly acquired characters. We shall see later, in the section dealing with the relation of dissociation to the mutation theories regarding bacteria, that Bordet also has raised the bacteriophage to a high eminence as an agent producing permanent mutations, and thus controlling the destiny of the species. But, as we shall also note later, both of these views are alike in their falsity. True bacterial mutations are, as yet, unknown, for we have been using false criteria for their attempted recognition."

-- p. 258 "Among the variants produced by the action of the bacteriophage in a sensitive culture some may still be sensitive (Bordet, Arkwright and others), some may be resistant without "carrying" the lytic agent, while still others are resistant and do "carry" the lytic agent (d'Herelle, Bordet, Gratia, Lisbonne and Carrere). The last are called lysogenic cultures and are able, through contact or through their filtrates, to precipitate lytic or growth-inhibitive reactions in normal, sensitive culture. Furthermore, in so doing, they cause the generation of more organisms like themselves, as well as more sensitives (perhaps) and more nonlysogenic resistants. D'Herelle and Hauduroy have made special provision for these lysogenic forms, either as organized cells of normal SR type, or as filtrable forms, "existing in symbiosis with the bacteriophage." While this conception of symbiosis between a bacteriophagic virus and filtrable forms of resistant bacteria requires a stretch of imagination for which many bacteriologists may not be qualified, I believe this last point mentioned above is an admission, on the part of d'Herelle, which represents the greatest, recent, single advance toward our understanding of the nature of the bacteriophage and of the probable mechanism of lytic action; and to this point I shall subsequently return." (see also pp. 225-296)

The above indented three pages are from Philip Hadley, MICROBIC DISSOCIATION, Journal of Infectious Diseases 1927, currently available in book form through Amazon.com and Kindle.

In 1942 Albert Tyler described "an approach to the problems of fertilizations from an immunological viewpoint", demonstrating and concluding "that certain specific substances are extractable from eggs and from sperm of various animals. ... [which] are evidently complementary proteins and they interact in a manner analogous to that of the complementary substances involved in the usual serological reactions, namely antigens and antibodies." [Tyler, A, 1942, West. J. Surg. 50:126.]

Tyler's discussion of auto-antibodies yields an interesting perspective on not only the nature of the bacteriophage, which it fascinatingly characterizes as "auto-antibody", but places the concept in a larger global context of the cell and life itself:

"If pathogenic bacteria likewise have complementary surface and subsurface substances, then the antibody to the organism should be obtainable simply by appropriate extraction of the sub-surface substance without the necessity of producing an antiserum in some laboratory animal. Since the antibodies that have protective value are those that react with the surface antigens, a complementary subsurface substance would be expected to give protection to an animal infected with the particular organism. To obtain the autoantibody it would, of course, be necessary first to remove the surface substance since the two would react and precipitate out if whole cells were subjected to extraction procedure. Likewise any other substances in the interior that were complementary to the sub-surface substance would have to be removed or the sub-surface substance would have to be present in excess. Preliminary tests have indicated the presence of auto-antibodies in blood cells but no direct evidence is as yet available concerning heir extraction from bacteria. The closest approach to this is perhaps contained in the work on bacteriophage by Burnet [GAC Gough and FM Burnet, J.Path.Bac. 38:301, 1934] who showed the phage to be complementary to the coat of the susceptible bacterium from within which the phage is in turn obtained. Phage, then, may be regarded as an auto-antibody.

The findings of two mutually complementary substances leads to the expectation that more may be found deeper within the cell. This view may then be expanded into a general theory of cell structure; namely, that a cell is a mosaic of substances that are mutually complementary (i.e., capable of combining with one another in the manner of antigen and antibody) substances which are actually in combination with one another in the regions where they adjoin The compounds would be represented by the various membranes, such as the cell membrane, nuclear membrane, nucleolar membrane, vacuolar membrane, etc., which in turn keep the complementary substances apart."

This is certainly fascinating "food for thought", and might be kept in mind throughout as we glimpse into an aspect of what might open to a whole new world of biologic understanding. We find the work of Tyler with particular relevance to the works of Pauling and Rosenow on the nature of antibodies and the development of artificial antibody, identical in all ways to natural antibody. With the work of Rosenow, this advanced to his miraculous successes with artificial antibody, proven as therapeutic in 40 cases of otherwise advancing and untreatable poliomyelitis in 1946 (reported by Rappaport in 1948) and 1948 (reported in 1954), and described by Rosenow as useful therapeutically in multiple sclerosis and a range of other chronic conditions as late as 1957 and 1958. Reprints of these articles from 1954, 1957 and 1958 are provided in the book, REFERENCE MANUAL ROSENOW ETAL, S. H. Shakman, InstituteOfScience.com, available through Amazon.com and Kindle books; articles are also available in various medical libraries.

Rappaport, Benjamin, *Quarterly Bulletin, Northwestern University Medical School* **28**, 1954, p. 57-60; Rosenow, E.C., Ohio M.M., 53(7), July 1957, p. 783-5.; Am. Practitioner and Digest of Treatment (Philadelphia), 9(5), May 1958, p. 755-761.

But retracing a bit to Tyler's discussion of the nature of the cell, and compatibility and reactions of its component parts, there is apparently an intimate relation between bacteriophage (i.e., Twort-d'Herelle) and antibody (as commonly regarded and studied relative to antigen), i.e., that they are not separate concepts which both may have a beneficial effect relative to pathogens, but rather that there is an intimate relation, even oneness, between the underlying beneficial actions of these otherwise-considered-separate entities. Further, notwithstanding the wonderful clinical results reported by Rappaport with Rosenow's artificial antibody (poliomyelitis) and Rosenow himself (in various disease conditions), the simplicity and thus promise of "bacteriophage" or "transmissible bacterial autolysis" seems potentially even greater.

Beyond its value as an historical review and documentation of work as of the 1920s and 1930s, this 3-part series of the works of Hadley and associates provides a consistent multi-hundred page comprehensive analysis of the phenomena of microbic dissociation, bacteriophage, bacteria versus viruses, etc. Rather than "reinventing the wheel" as we so often seem to want to do, those who would now wish to go further with the issue could be well advised to use this as at least a good starting point. As Edmund Burke (1729-1797) long ago stated, "Those who don't know history are destined to repeat it." So it seems that a lot of ground might be covered in a minimum of time, perhaps not to solve all the attendant problems, but at least to bring us closer and more expeditiously to a solution.

The main reason for bringing these works out in book form is simply that they do not otherwise exist in a readily-accessible manner. Yes, the original journal articles can be accessed through many, primarily medical, library facilities, but for proper study, markup, digestion and review, a hard copy seems helpful or even necessary. And if this in some small way helps bring this perspective to broader attention, utilization and hopefully successful advancement, this writer is forever humbled and grateful for the honor of having been its agent.

S. H. Shakman
Executive Director
Institute Of Science, Santa Monica, CA, USA
InstituteOfScience.com, I-o-S.org
July 24, 2013

THE TWORT–D'HERELLE PHENOMENON

A CRITICAL REVIEW AND PRESENTATION OF A NEW CONCEPTION (HOMOGAMIC THEORY) OF BACTERIOPHAGE ACTION

PHILIP HADLEY

From the Hygienic Laboratory of the University of Michigan, Ann Arbor

OUTLINE

Received for publication, Oct. 18, 1927.

1. INTRODUCTION

In introducing the present work, I will first say a word concerning reviews in general. For many years I have believed that bacteriology, more than other fields of the biologic science, has suffered—in fact, is suffering—from the lack of suitably frequent correlation between many of its leading lines of advance. Important salients, thrust into the territory of the unknown, may easily lose their new found advantage if left too long destitute of liaison with supporting movements elsewhere along the front. One may contend that every captain and lieutenant in the army of research should be his own liaison officer—should establish and maintain his own connections with the rear, as well as with other salients. But, although of unquestioned value when possible, this task is becoming more and more difficult, and perhaps impossible. The field of operations, even in a single sector of science, has become too extensive, and the modes of attack too diversified. Therefore, reviews and correlations of scientific literature are becoming each year of greater necessity, and journals which devote themselves to such reviews, more and more valuable in the furtherance of research.

Thus, if any explanation is required for the preparation of this critical review on the bacteriophage, it is that several aspects of the problem of the bacteriophage demand correlation. Moreover, it seems desirable to present certain phases of older studies with greater, or with different, emphasis than that characterizing d'Herelle's latest (1926) comprehensive work; also, to point out several striking misconceptions which his various contributions seem either to have established or upheld. In addition, since the field of research on the bacteriophage has broadened somewhat since 1925, it is worth while, perhaps, to present certain newer matters of interest and to correlate them with facts acquired at an earlier date.

In the following presentation of certain aspects of the problem it will probably seem to many readers that I am departing considerably from the usual limitations of the subject. In such a case it will be because my conception of the problem of transmissible bacterial autolysis is broader than d'Herelle has been willing to admit. For him, whatever phenomena depart in any way from the classic autolytic reaction of the Shiga dysentery bacillus are relegated to other fields of autolytic change. But those who are acquainted with my earlier review of microbic dissociation will readily appreciate some of the grounds on which I have reached the conclusion that the Twort-d'Herelle phenomenon is merely one phase of a much greater problem which must be explained—namely, dissociative behavior among bacteria at large. Thus, although I cannot regard the phenomenon of the bacteriophage as comprising so narrow a field of reaction as d'Herelle would have us believe, I shall, in the present treatment, attempt to avoid, so far as possible, repetition of certain aspects of the problem considered in my earlier publication.

Finally, this review will not be concerned with all recent literature on the bacteriophage. Much has been written about the phenomenon which, though interesting in its bearing on some phase of the subject, does not bring one any closer to an understanding of the nature of the phenomenon or of its underlying mechanism. Such contributions, while often valuable, will be omitted for economy of space, and because the present work is mainly concerned with the nature and the mechanism of the bacteriophagic reaction. In view, however, of the important relationship of the bacteriophage to problems of immunity and preventive medicine, an exception will be made in dealing with these aspects of the subject.

2. THE ORIGIN OR "SOURCE" OF THE LYTIC AGENT

Since d'Herelle's * first isolation of the bacteriophage of the dysentery bacillus from human fecal matter in 1916, there has been a tendency to regard the lytic agent as intimately associated with the intestinal

---

* On this and all following pages of this work the reference numbers will not commonly be given after the names of authors if they are included in the reference list heading the section. If the citation is not included in the reference list, it will be placed after the author's name in the body of the work.

tract, or with substances which have opportunity to receive fecal contamination. In 1920, Bordet and Ciuca isolated the agent from the peritoneal cavity of guinea-pigs infected with Bacillus coli; they believed that its presence was due to the influence of the leukocytes; d'Herelle explained that its presence was attributable to bacteriophage that had migrated through the intestinal wall into the peritoneal cavity. As time has passed, however, the bacteriophage has been "isolated" from many other "sources," some of which would seem to preclude the possibility of intestinal contamination. Moreover, as will be seen, it has been generated artificially from normal, sensitive cultures by resort to a number of simple laboratory procedures. These results discredit the usual interpretation of many experiments in which the bacteriophage is alleged to have been "isolated" from one source or another.

REFERENCES

Glycerolated Vaccine Virus: Twort;[322] Gratia.[152, 154]

Intestinal Tract: d'Herelle;[185] d'Herelle[190] (technic); Bordet and Ciuca[44] (supported by B. Coli); Sonnenschein[315] (extensive study).

Peritoneal Cavity: Bordet and Ciuca[41,42] (for B. coli); Wollenstein[344] (confirmed Bordet and Ciuca);[344] d'Herelle[189] (failed to confirm Bordet and Ciuca; Bordet[38] (confirmed earlier results); Meissner[256] (obtained agent after injecting cholera vibrios and cholera immune serum); Petrovanu[277] (results not confirmed).

Intestinal Mucosa: Petrovanu[276] (V. cholerae); Collins[60] (B. cholerae suis); Kuttner[223] (B. typhus); Jonesco-Mihaisti[202] (Shiga); Borchardt[34] (duodenal mucosa of cat, Shiga); Prausnitz and van de Reis[287] (human small intestine); Petrovanu[279] (small intestines of normal and inoculated rabbits); Pyle[291] (mucosa of domestic fowl).

Bladder (of normal guinea-pigs—B. Coli): Larkum.[230]

Appendix: McKinley[249] (for Shiga cultures).

Spleen: Villazon[334] (agent for rat pest); Kuttner (B. typhosus).

Liver: Kuttner[223] (guinea-pig and rabbit); Pyle[291] (from fowl, for Bact. pullorum).

Kidney: McKinley[249] (of guinea-pig, for Shiga).

Ovaries: Pyle[291] (of fowl, for Bact. pullorum).

Empyema Fluid: McKinley[249] (human, for Shiga).

Salivary Gland: McKinley[249] (human, for Shiga).

Prostate Gland, also Prostatic Fluid: McKinley[249] (for Shiga).

Mouth: Sierakowsky and Zajdel[312] (for B. acidophilus).

Tonsils: Petrovanu[276 a] (for B. coli in case of angina).

Pus: Callow;[52] Gratia;[152, 154] Lisbonne, Boulet and Carrère;[233] d'Herelle;[189] Kropveld.[207]

Nursling Stools: Suramyi[320] (normal and pathologic); Kropveld;[220] Pierrot[282] (in new-born).

Peptone Solution: Hadley and Klimek[170] (for Shiga and B. coli).

Fresh Pancreas: Borchardt;[34,35] Keller;[212] Arnold and Weiss;[16] Hoder and Suzuki.[298]

Pancreatin (Commercial or Pancreas): Borchardt [34, 35] (failure with Handel's preparations); Bachmann and Aquino; [21] McKinley; [248] Pico; [290] Hoder and Suzuki; [196] Hadley and Klimek [179] (extracts of several American preparations, sterilized and unsterilized); Fränkl and Schultz [357] (from hog pancreas).

Papain: Pico; [290] McKinley; [248] Combiesco [91] (unsuccessful after forty passages).

Trypsin: Combiesco; [91] Jötten; [204] Pico; [290] Borchardt [34, 35] (from fresh pancreas, but no success with commercial preparations); McKinley [248] (unsuccessful); Pondman; [283] Keller; [212] Wollman [331, 335] (unsuccessful with several samples); Hadley and Klimek [170] (successful for Shiga); Wolff [248] (successful with six of seven preparations of trypsin); Hadley and Klimek [179] (successful for Shiga and B. coli with one sample sterilized at 120 C.).

Pepsin: McKinley [248] (unsuccessful).

Insulin: McKinley [248] (unsuccessful).

Leukocytes: Bordet and Ciuca [61] (instrumental in modifying the bacteria); d'Herelle [182] (disclaims this view); Lisbonne, Boulet and Carrère [233] (obtained from normal leukocytes in vitro); Marcuse [244] (leukocytes not necessary for its generation); Fränkl and Schultz [357] (by action of leukocytes in vitro).

Bacterial Antagonism: Kuhn (1919) (B. typhi murium vs. Metchnikoff's vibrio); Lisbonne and Carrère [234] (B. coli vs. the Shiga bacillus), (B. coli vs. the staphylococcus); Beckerish and Hauduroy [50] (dispute the conclusions of Lisbonne and Carrère); Farby and van Beneden [122] (B. coli vs. white staphylococcus), (B. prodigiosus and B. coli).

Bacterial Associations: Flu [381] (negative with Shiga bacillus and B. pyocyaneus); Pondman [283] (pseudodysentery Y and B. pyocaneus).

Pyocyanase: Pondman [282] (negative results with Shiga culture, but positive with Hiss "Y14").

Lithium Chloride: Kuhn [221] (generation of bacteriophage by culture growth on l. c. agar).

Methyl Violet: Pondman [283] (spontaneous lysis through addition of this substance to medium).

Milk: Sonnenschein [335] (from centrifugal slime and filter residuum—for Flexner cultures only).

Cheese: Sonnenschein [335] (bacteriophage for B. proteus, HX19).

Nasal Secretions: Sonnenschein [335] (for capsulated bacteria in cases of rhinitis atrophicans).

Water (or Sewage): Dumas, [109] Arloing, Sempé and Chavanne [12] (rivers of France); Manoliu and Costin, [242] Zdansky [345] (various); Monteiro [257] (of San Paulo); Sangiorgi and Vercellana [299] (of Italy); Nakashima [260] (of Germany); Arloing and Sempé [12] (foreign rivers and Atlantic ocean); Collins [90] (Michigan rivers); Arnold [14 b] (Illinois); Caldwell [81] (Texas); Clark and Clark [80] (river sludge); Hadley and Klimek [179] (sterile, double-distilled water).

Nodules of Leguminous Plants: Gerretsen, Gryns, Sack and Söhnigen [196] (bacteriophage for Ps. radicicola).

Rot of Carrots: Coons and Kotilia [92] (bacteriophage for B. carotovorus).

Spontaneous Origin in Cultures or Culture Filtrates (exclusive of recognized "lysogenic" cultures): Bail [26 a] (in three strains of Flexner bacillus).

Béguet [51] (successful by adding sensitive cultures of the Shiga bacillus to filtrates of naturally resistant cultures).

Clark and Clark [80] (not successful with S. mucosus).

Dewey and Green [106] (from cultures of typhoid which had proved resistant to heterologous lytic principles).

Euguchi [119] (from filtered dysentery cultures).

Fejgin [120] (from B. proteus).

Fejgin [124 a] (from cultures of Shiga dysentery bacillus).

Fejgin [127] (from B. diphtheriae).

Flu [143] (one culture among 100 cholera strains showed spontaneous autolysis).

Gaté and Gardère [130] (pyocyaneous lytic agent from old cultures).

Gildemeister and Herzberg [135] (from old dysentery cultures).

Hadley [163] (from fresh cultures of B. pyocyaneus).

Hadley and Klimek [170] (from old broth cultures of B. coli and Shiga, filtered repeatedly).

Hadley and Klimek [170] (bacteriophage for Shiga cultures produced experimentally through enforced microbic dissociation).

Israelsky [200] (unsuccessful with B. tumefaciens).

Jötten [204] (successful through self-digestion of heavy culture suspensions after ten days' incubation; coli, typhoid, dysentery, paratyphoid).

Kline [261] (from fourteen of twenty-one cultures of intestinal bacteria).

Kuhn [221 a] (from bacterial antagonism—B. typhi murium vs. Metchnikoff's vibrio).

Kuhn [221 b] (obtained lytic filtrates from washings of seven-hour cultures grown on agar containing lithium chloride).

Kuhn [221, 222] (correlation between generation of bacteriophage and presence of Pettenkofer A bodies in cultures).

Kuttner [235] (from old broth cultures of B. typhosus).

Lisbonne and Carrere [234] (in B. coli, by associations and by use of leukocytes in vitro).

Manniger [261] (culture changes characteristic of lytic action in B. coli cultures).

Nobechi [262] (by repeated filtrations of the cholera vibrio grown in their own filtrates).

Otto and Munter [280] (from old cultures).

Seiffert [331] (from old Shiga cultures).

Wollman [325] (not successful with Shiga cultures).

DISCUSSION

Source.—The citations made in the foregoing section would seem to imply that the so-called "sources" of the bacteriophage are sufficiently various. They apparently support d'Herelle's view that the agent is to be found wherever there is opportunity for fecal contamination; and that, from such sources, it can be isolated through the selective action of the bacteria. On this subject, however, the opinions of bacteriologists have, from the first, been divided. Some have sought for the source outside the bacterial cell, while others have held that the bacteriophage is always derived from the bacteria themselves. The latter view is naturally held by those who support Bordet's theory regarding the nature of the lytic agent. In every specific instance in which the agent has been alleged to arise in the bacteria themselves d'Herelle has met these threatening observations with the statement that the cultures employed have been, at some time or other, "contaminated with the bacteriophage."

In examining the so-called sources, the question, therefore, arises, "Does the lytic agent exist as such, and in a preformed state, in these substances, or do these sources merely represent certain influences which, when brought to bear on the bacterial culture, force it to generate the lytic agent de novo?" As I have already pointed out with Klimek,[170] the definite proof of the origin of the lytic principle within the bacterial cell is of great significance, for it constitutes the one possible demonstration that can eliminate from the field of controversy d'Herelle's theory of a filtrable virus of the bacteriophage.

As to glycerolated vaccine, from which Twort first, in 1915, isolated his lysing cultures of staphylococcus, there is slight basis for an opinion as to the presence of the lytic agent in this substance. So far as I am aware, no tests have been made that might reveal the agent in the vaccine separated from the organisms whose lysis it effected. It first appeared already bound to the bacterial substratum. Subsequently, it could, as is known, be obtained free in the filtrates. On the whole, it is possible to suspect that the vaccine, some of its components or the manner of its treatment or preservation served as an "influence" on the contaminating staphylococcus organisms.

Regarding the intestinal content, it is reasonable to believe that the lytic agent often exists in the intestines in a preformed state; and, even when one considers the intestinal or duodenal mucosa, there is the possibility that the principle might exist there as a result of contact with the intestinal contents. I do not recall, however, that any one has attempted to produce lytic areas directly through the use of fresh and unmodified filtrates of the intestinal contents or of extracts of mucosa. It is also possible to conceive that the lytic agent might make its way from the intestinal reservoir to the peritoneal cavity, to the blood stream and to the organs and glands of the body, so that it might be isolated from these sources, as has been alleged by many investigators. It is even possible to assume that its presence in the glandular tissues of animals might account for its alleged abundance in such preparations as pancreatin and all the commercial enzyme products. On the other hand, we know from the studies of Krestownikowa and Gubin [219] and others that bacteriophage artificially administered remains but a short time in the blood or other tissues (less than $8\frac{1}{2}$ hours). Its migration from the intestines into tissues is improbable.

By the use of such products as those mentioned the lytic principle can undoubtedly be quickly obtained for a variety of bacteria, so that Hoder and Suzuki have even recommended such a "source" (commercial pancreatin) as preferable to intestinal material. I believe that the observation, if not the mode of explanation, of these authors is correct, and I have reached this conclusion after a large number of experiments performed with my student, Klimek, on samples of American pancreatins. But it also appears from our studies, which will be reported in detail at a later date, that all preparations are by no means of the same value in determining the generation of the lytic agent. In our experience the Squibb product was superior in bacteriophage-generating power to two others examined. Although Hoder and Suzuki were inclined to the view that the bacteriophage was present in abundance in their pancreatin preparation (Pancreon), and are thus in agreement with several other investigators (as L. K. Wolff), a few workers, for example, Keller (1924), have presented a different point of view. In trypsin, for instance, Keller has seen the possibility of a principle not itself lytic, but "liberating." He was unable, however, to demonstrate the important point at issue.

Hoder and Suzuki approached the same possibility and even went so far as to test the power of fresh pancreatin or pancreatic extracts to register directly (in the form of lytic areas) on agar plates seeded with sensitive culture. Since they did not succeed in this attempt except in one or two questionable instances, they concluded that, although the lytic agent existed in abundance in pancreatic tissue and in pancreatin, it was so closely "bound" to the organic constituents that it was unable to register directly on plates by the production of the customary lytic areas.

In the experiments which I have conducted with Klimek, however, in which we attempted to cause the binding of Squibb pancreatin with lytic principle which had already been produced by the use of other samples of the same product, we could obtain no indication of the occurrence of this union. The amount of the lytic agent was never reduced nor lessened in its activity for the sensitive test-culture.

Taken as a whole, our experiments on the generation of the bacteriophage by the use of extracts of pancreatin have conclusively demonstrated the fallacy of the view that pancreatin contains the agent in a preformed state. After we had demonstrated that the extracts readily caused the generation of the principle for Shiga, for B. typhosus and, less easily, for B. coli, we heated the pancreatin extracts in the autoclave at 110 C. for thirty minutes, or at 120 C. for twenty minutes. When these extracts were employed they were still able to cause the generation of the principle in the sensitive cultures, although more slowly than in the case of the unheated extracts. It may, therefore, be concluded that it is not the active enzymes that are responsible for the appearance of the bacteriophage in the cultures. The same result was obtained through the use of sterilized peptone solution (Parke Davis Company bacteriologic peptone), by inactivated trypsin solutions, and, finally, it may be added, by sterile, double-distilled water.

The same situation exists in the case of sewage filtrates as appears from the experiments which I have performed with Kiesewetter. Here, if quite fresh, no indication of a preformed lytic agent can be detected by plating methods, although a small amount of the sewage filtrate introduced into broth results in the generation of abundant lytic agent which then registers on agar plates in the usual manner. One may be inclined to say that the lytic corpuscles exist in sewage in great number, and so one would expect. But what shall we conclude when we find that sewage filtrates, heated under steam pressure at 120 C. for twenty minutes, also serve, after ten to twelve serial inoculations and filtrations, to generate the principle? Perhaps the only answer that can be given to this question at present is, that the corpuscles exist in sewage, but not in a form that permits them to make a direct registration.

We cannot suspect the presence of a lytic agent endowed with powers of life existing in sterilized sewage water, sterilized pancreatin extracts, sterilized peptone solutions or sterilized double-distilled water. Yet each of these agents serves to generate the lytic agent in broth cultures. The tests with the sterile pancreatin extracts particularly serve a double purpose. They demonstrate that, if the agent exists preformed, as such, in the samples of pancreatin, it is not a living thing in the d'Herelle sense; furthermore, that, if the action of the pancreatin is merely one causing the "liberation" of the bacteriophage from the bacteria themselves, it is not due to the active enzymes of this substance. What has been said of pancreatin can probably be said with equal truth of the majority of the other "substances" from which the lytic agent is alleged to have been isolated.

Spontaneous Origin of the Bacteriophage.—I come now to those instances in which the lytic principle has seemed to appear in bacterial cultures spontaneously, so-to-speak; or in which they have been submitted to no other treatment than repeated growth through their own filtrates. It is these instances, particularly, that afford the clearest evidence in favor of the autogenic nature of the bacteriophage.

Beginning with the observations of Bail and of Otto and Munter on the spontaneous appearance of the bacteriophage in apparently normal cultures, the number of such reports has rapidly increased, and there is now clear evidence that it is unnecessary to resort to the use of any foreign substance in order to obtain the lytic agent for a given culture. Indeed, Otto and Munter have long disparaged the common practice of searching for "Fundorten" of the bacteriophage. I have shown with Klimek, and also with Kiesewetter, that a strong and typical lytic agent can be generated in a tube of broth merely by repeated growth and filtration through a series of 14 to 20 operations, and probably less. The same has recently been reported by Nobechi for the cholera vibrio, apparently with still less difficulty; also by Kline for intestinal bacteria, by still simpler means. Failures to accomplish this end, as reported by Wollman, the Clarks, Israelsky, Sonnenschein and others, cannot at present be explained, but the number of positive instances is sufficient to carry the point. I believe there does not exist such a thing as is termed by d'Herelle "an ultrapure strain" of bacteria; at least, no strain is at all times "ultrapure."

In view, however, of d'Herelle's criticism of all such cases, one related matter must be considered in its bearing on the spontaneous appearance of the bacteriophage from the culture itself, and this is the question of the so-called lysogenic cultures. In d'Herelle's explanation of spontaneous origin he is naturally driven to postulate a previous "contamination" with the bacteriophagic virus; and cultures so contaminated are spoken of by him as "mixed." The bacteria and the bacteriophage are living in a state of symbiosis. Although I regard this view as far from the truth, it must be admitted that there exists a form of culture which, superficially considered, seems to offer grounds for d'Herelle's view, and this is the lysogenic culture, which is capable of generating the agent, immediately and without manipulation, and, therefore, is capable of producing the lytic reaction in other sensitive cultures, but which, itself, is resistant to the principle that it generates. Such a culture presents all the characteristics one would expect to observe in a culture "infected" with bacteriophagic ultravirus, but one which has become partially resistant. Indeed, as Bordet [40] has shown, such a type of culture may be produced artificially by submitting a normal, sensitive culture to the action of a weak lytic principle, and then selecting certain of the resistant secondaries. But this form of growth is quite different from that in which the spontaneous origin of the lytic agent has usually been observed. In my own experiments the strains of departure have been proved to be normal, smooth, S type cultures, acutely sensitive to the bacteriophage developed from them.

Indeed, the original culture was often used to test the fact that a lytic principle had been generated. There is thus a difference between the discovery of a bacteriophage in lysogenic cultures and the finding of one in sensitive cultures, which for years have given a record of normal behavior. If the bacteriophage is discovered to have arisen in cultures of the latter sort, the circumstance must be accepted as indicating a spontaneous origin. This undoubtedly holds for the greater number of recent instances in which the sudden appearance of the lytic agent has been noted. On the other hand, I believe that all "normal," sensitive cultures (S cyclostage) are potentially lysogenic; also many R cultures.

Conclusions.—The time has arrived in studies on the bacteriophage when one should be little concerned with the so-called source or origin of the lytic principle outside the bacterial cell. We must recognize many, and perhaps nearly all, of the alleged "sources" as representing nothing more than "liberating influences" which can force the culture, if fully sensitive, into a new form of growth in which the lytic agent is generated de novo from the cells of that culture. In studies on the nature of the principle we are, therefore, confronted with the necessity of differentiating between the indefinite, liberating stimuli which set into operation the machinery of bacteriophage-generation, and the definite active agent which is able to maintain the lytic process in the culture itself or in indefinite series. The exact nature of the liberating mechanism and its mode of influence on the young culture thus becomes a problem approaching in importance that of the nature of the lytic units themselves.

3. THE BACTERIOPHAGE UNIT AND SOME OF ITS PHYSICAL CHARACTERISTICS

The ultimate form or structure of the bacteriophage possesses significance in suggesting its actual nature and mechanism of action. Many bacteriologists have come to accept the evidence early furnished by d'Herelle as demonstrating that the agent is corpuscular; that it is a physical unit, as opposed to a substance in solution. Some have regarded it as a bacterial "fragment." On the other hand, it has been alleged that the ability to register in the form of plaques or lytic areas is not in itself sufficient evidence to establish the view of a physical unity for the bacteriophage. Bordet and Ciuca [40] have, from the beginning of their studies, maintained an hypothesis which manages to escape the implications inherent in d'Herelle's corpuscular theory. This involves their notion of a "nutritive vitiation" which, in a sense, becomes hereditary in the culture, being transmitted through the mechanism of a substance of chemical nature generated in, and liberated by, the bacteria. In this section of the paper, some of the details of this theory will be examined, together with the bearing of certain observations on filtrability, size, electrical charge, adsorption, elution, manner of lysis, etc., on the various conceptions of the nature of the lytic agent. The more specific consideration of the possible nature will, however, be reserved for another section.

REFERENCES

Corpuscular Nature: d'Herelle;[188] Wollman;[330, 335] Asheshov;[17] Collins;[40] Marshall;[246] Bronfenbrenner,[354 b, 354 c] and others.

Chemical Nature (non-living): Bordet and Ciuca;[41] Bordet,[40] and others.

Filtrability of Corpuscles: d'Herelle;[188, 193] Prausnitz;[284, 367] Maisin;[240] Wollman;[325] Biemond;[32] Stassano and Beaufort;[336] Wollman and Suarez;[349] Eliava and Suarez;[135] Kramer;[213] Bechold;[354] Levaditi and Nicolau;[231] Zinnser and Tang;[348] Wollman;[330] Fränkl and Schultz.[357]

Size of Corpuscle: Prausnitz;[284, 367] von Angerer;[5] Levaditi and Nicolau;[231] Wollman and Suarez;[349] Marshall;[246] Zinsser and Tang;[368] Wollman;[330] Bechold;[354] Fränkl and Schultz.[357]

Diffusion: d'Herelle;[188] Arnold;[16] Jötten;[204] Prausnitz and Firle;[284] Flu;[131] Marshall;[246] Callow;[82] Weiss.[320]

Electrical Charge: Adsorption: d'Herelle;[193] Meissner;[256] Arloing, Langeron and Sempé;[11] Prausnitz and Firle;[286] Collins;[60] Koch;[214] Gildemeister and Herzberg;[188] Seiffert;[310 a] Bechold;[354] de Poorter and Maisin;[306] Marshall;[246] Jaumain and Meullman[301 a] Clark and Clark;[50] Kramer;[218] Callow;[83] Fränkl and Schultz;[357] Kramer.[215]

Cataphoresis: Koch.[214]

Elution: de Neckar;[262] Gildemeister and Herzberg;[188] Callow.[83]

Lysis Under Microscope: d'Herelle;[193] Wollman;[325] Costa Cruz;[90] Larkum;[227, 228] Kuhn;[221, 222] Béguet;[31] Bronfenbrenner;[354 a] Koch and Ziegenspeck.[216]

DISCUSSION

Corpuscular Nature.—The majority who have expressed an opinion on the subject have conceded that lytic areas on solid culture mediums represent the site of development of independent lytic units which have multiplied at the expense of the surrounding culture. The clearest evidence for this conclusion has been supplied by d'Herelle and by Wollman.[336] Against this conception, however, Bordet and Ciuca brought their view that, since the lytic agent is a chemical substance in solution, it cannot register as in the picture of single, physical units; therefore, the mechanism of plaque formation must lie in the behavior of the bacterial cells. They believed that, if the bacteriophage is present in strong concentration, it affects a considerable number of especially sensitive cells in the substratum, precipitating them into lysis. If the lytic substance is, however, present in weaker concentration, it affects a smaller number of sensitive cells in the substratum. Thus, for lytic filtrates of different concentrations, there will be different numbers of sensitive cells capable of receiving the influence. Since the lysis of these cells (and the subsequent spreading of the lysin to surrounding cells) is responsible for the appearance of the lytic areas, according to Bordet, the lytic filtrate of greatest concentration will determine the greatest number of areas, while a less concentrated filtrate will determine a smaller number.

If Bordet's conception were true, one would anticipate that the number of plaques produced by a given lytic filtrate on a sensitive culture would vary with the number of bacteria present, as well as with the dilution of the lytic agent. The experiments devised by d'Herelle to test this point had the merit of demonstrating that the number of lytic areas is a function of the amount or concentration of the lytic filtrate, and not of the bacterial culture. This was also the opinion of Collins [90] from experiments performed in this laboratory, and of Marshall [246] who repeated carefully the experiments of both d'Herelle and Bordet. The same conclusion has been reached by Asheshov, Wollman and many others. Wollman's experiments reported in 1927 are particularly convincing. Slight inconsistencies have sometimes been pointed out, but, I believe, the truth of d'Herelle's original statement is now fully established. The number of lytic areas is determined, roughly at least, by the amount or concentration of the lytic filtrate, and not in an appreciable degree by the number of bacterial cells in the culture. As we shall observe later, however, it may be modified by the kind of bacteria in the culture. It has been shown, moreover, that when the degree of resistance of a sensitive culture toward the bacteriophage is increased, the first indication is not always a diminution in their number but often a lessening in their size. This phase of the subject will be considered further in another section of this paper (section 7).

Filtrability and Electrical Charge.—The filtrability of the bacteriophage through candles of porcelain and of diatomaceous earth is too well known to demand comment, aside from the effect of charge, which will be considered presently. But, in addition, a considerable study has been made of filtration through membranes of gelatin or collodion. Prausnitz was one of the first to suggest the probable size of the bacteriophagic corpuscle as a result of filtration through gelatin membranes of graded porosity (de Haen), and placed the dimensions as about 20 millimicrons, very near the assumed size of the micellae of collargol, and much larger than trypsin molecules. Maisin did not succeed in observing the dialysis of the agent through the membranes of Abderhalden. Wollman, however, was able to demonstrate passage through collodion membranes if they were not too dense. Similar results were obtained by Biemond who concluded, probably incorrectly, that the size of the lytic unit was smaller than the molecule of serum albumin. Stassano and Beaufort showed that collodion filters that held back strychnine nitrate (molecular weight 397) and certain ferments, such as the peroxidase and reductase of milk, let the bacteriophage pass. Wollman and Suarez used filters of different porosities and found that the lytic agent easily passed membranes of 7% collodion which retained hemoglobin and serum proteins. Eight % membranes yielded a less active filtrate and a 10% membrane permitted only the occasional passage of a corpuscle. The fact, however, that the agent passed filters impermeable to serum proteins suggested to them that the bacteriophage is related to substances different from the coagulable proteins. They ascertained that a filtrate through a 7% membrane, yielding several million units per drop, gave a negative reaction to the nitric acid test for protein. The residue left on the surface of a 10% membrane gave a positive reaction to the protein test, but these workers were not able to sensitize guinea-pigs with this material. These authors concluded that the lytic corpuscles are variable in size and that this circumstance can explain certain divergent results. Eliava and Suarez, realizing that conclusions on filtrability are valueless unless the work is done under strictly determined conditions, sought to ascertain the value of different factors coming into play, namely, adsorption, charge, etc. They employed the Bechold-König assembly, with porcelain cups coated with 8% collodion and equipped with an agitator turned by a motor just above the membrane. The filtration was conducted by aspiration at 400 mm. Using a Shiga principle they found that it filtered readily unless the membrane was made positive; then absorption was great and filtration rendered difficult. This naturally suggested a negative charge carried by the corpuscles. Even with negative membranes, however, the filtrate was about 200 times less active than the original material.

Bechold, using his own filter assembly, showed that the albumin molecule passed filters that held back the bacteriophage lysins. From this and other tests, he therefore concluded that the bacteriophage units were larger than the molecules of albumin (from 4 to 10 millimicrons) and were larger than the enzyme molecule. From other tests he regarded them as also larger than the colloidal particles of collargol (estimated at about 35 millimicrons). He believed with Otto and Munter [304] that the bacteriophage is a disintegration product of the bacterial cells in which it arises. Fränkl and Schultz observed that the lytic particles passed the deHaen membranes 1 to 6, while congo-red was held back by membrane 6. They also noted that the bacteriophage was never adsorbed by substances possessing a positive charge.

D'Herelle believed that the principle carried a negative charge, but this view has been opposed also by Seiffert and by Gildemeister and Herzberg who showed that it is easy to adsorb the lysin on electronegative colloids, but more difficult to adsorb it on colloids with a neutral or positive charge. Koch believed that lytic units carried a positive charge. His conclusion was based on tests of cataphoresis. He observed, from the use of special apparatus involving unpolarizable electrodes, that the strength of the bacteriophage suspension increased at the cathode and decreased at the anode. Prausnitz and Firle concluded that the easy absorption on negative colloids indicated a positive charge. Meissner had shown that when the lytic corpuscles were brought into contact with positively charged bacteria the lytic agent did not weaken. This author also observed that a culture in the sensitive state bound the lytic corpuscles, while in the resistant state it did not. She therefore came to the conclusion that the binding between the bacteriophage and the bacteria does not necessarily depend on absorption in any way, but represents a chemical binding; and she believed that this reaction was analogous to the sort of binding observed in serologic reactions, as in agglutination, for example. As will be indicated later, it may not be necessary to assume that the binding is dependent on the phenomena of adsorption. It is permissible to believe, however, that a chemical attraction exists. To the foregoing it should be added that recently Kramer observed that certain substances such as the bacteriophage (antistaphylococcus), vaccine virus and rabies virus, which easily pass Berkefeld candles (commonly possessing a negative charge) were held back by filters composed of plaster of Paris plus magnesium oxide (possessing a positive charge).

Size of the Lytic Corpuscle.—Although d'Herelle from his earlier experiments in ultrafiltration had concluded that the bacteriophage unit possessed a size about equal to the micella of serum albumin, Prausnitz was the first to calculate the dimensions as about $20\mu\mu$ as a result of gelatin filtration tests. Von Anger, by the use of refraction methods, was led to place the size as about $30\mu\mu$. Biemond, by filtration tests through collodion, found that the bacteriophage did not pass filters that permitted the filtration of serum albumin and hemoglobin. Stassano and Beaufort saw the corpuscles pass filters that held back nitrate of strychnine. Wollman and Suarez, using acetic collodion, found that the bacteriophage passed filters that held back the serum proteins. It is to be noted, however, that while the tests of the first portion (15 cc.) coming through the filters were negative for bacteriophage, the two hour filtrate revealed bacteriophage. Levaditi and Nicolau observed that the filtration of the bacteriophage was obtained under the same conditions that permitted the passage of the virus of rabies, vaccinia and encephalitis (herpes). Marshall has recorded the size of the lytic corpuscle as between 10 and 100$\mu\mu$. His conclusions were based on experiments involving the number of lytic areas produced on agar plates. In these tests the number of lytic units present in a definite quantity of agar was known, and the number appearing on the surface of the agar (registering in the form of lytic colonies) was regarded as a function of the concentration.

Bechold, from his own filtration experiments, concluded that the lytic units were larger than molecules of albumin (from 4 to 10$\mu\mu$); also larger than enzyme molecules and the micellae of collargol. Fränkl and Schultz, from their tests with the deHaen membranes, believed that the bacteriophage possessed a size below that previously recognized as the limit of known living things.

Although the actual, or even approximate size of the bacteriophagic units and particles of the filtrable viruses must still be regarded as unknown, there remains the possibility of ascertaining the relative size of these elements among themselves and of making comparisons between them and the colloidal particles of other known organic or other chemical substances. In default of more accurate information such data possesses considerable value. An attempt to obtain such information has recently been made by Zinsser and Tang, who made use of a specially devised filtration technic, involving collodion membranes and affording adequate control measures. They made comparisons of the filtrability, through different fractions of the same stock of membrane, of crystallized egg albumin, crystallized serum albumin, trypsin, collargol, casein, staphylococcus bacteriophage, Rous sarcoma virus, herpes virus and arsenic trisulphide. Briefly stated, the results indicated that the size of the bacteriophage unit corresponded closely with the size of the virus particles. It was, however, smaller than the micella of arsenic trisulphide, but larger than the molecules of egg albumin, serum albumin, trypsin, collargol or casein. These results would suggest a size for the lytic unit greater than $20\mu\mu$ but less than $100\mu\mu$. It is of considerable interest that the size of the bacteriophage corpuscle was found to agree closely with that of the particles of the two recognized viruses (Rous sarcoma and herpes). In a general way, the work of these investigators suggests a size for the bacteriophage somewhat greater than that indicated by earlier studies in this field.

Diffusibility of the Bacteriophage.—As noted by d'Herelle, contributions to the subject of the diffusibility of the principle have been made by Jötten, by Prausnitz and Firle and by Flu. While the first believed that diffusion occurred into 3 per cent agar, the latter observers concluded that diffusion through agar did not take place. D'Herelle concluded that such diffusion through agar was improbable. Marshall, however, employing his so-called "live agar" (agar containing a suspension of living bacteria), came to the conclusion that the principle diffused at least through 1 mm. of agar, causing the lysis of the bacteria below the agar surface. He also showed that it readily passed through certain collodion membranes without pressure. Evidence of diffusion through agar is also found in the work of Arnold.

Regardless of the ability of the bacteriophage to penetrate agar, however, I was able to demonstrate several years ago, in some still unpublished experiments, that the principle can pass over stretches of sterile agar in passing from one mass of culture to another. This I proved by two methods: by use of colonies and of streaks on agar plates. In the first case, the following method was employed. In the center of a sterile and partly dried agar plate was started, by drop inoculation from broth culture, a colony of the sensitive culture (a coli-like organism from river water). As soon as it had dried, the center of the broth-drop was inoculated with a trace of bacteriophage, and in such a way that the latter would spread slowly through the broth-drop area from its center as the colony (lytic) developed. Next, other colonies of culture were planted, by drop inoculations, at varying distances from the central colony. The distances varied from 1 to 10 mm. Naturally, through the subsequent growth of the colonies, these distances were later diminished. The object was to ascertain whether the bacteriophage would travel across the bare agar from the central lytic colony to one or more of the surrounding colonies; and, if so, how far it would travel. The results showed, in brief, that the principle would pass in this way over such areas of sterile agar and attack the neighboring colonies on the margins nearest the central colony, but only if the distance was not greater than 2 to 3 mm. Passages of this distance, however, were not common, although passages of from 1.2 to 2 mm. were often observed. Under these conditions the neighboring colonies would undergo lysis on the side nearest the central lytic colony, and yield at this site, not only the bacteriophage, but typical secondary colonies together with the usual gray (a) zone. In some cases the entire colony would undergo lysis as a result of the passage of the lytic agent through the colony, as in Twort's original case.

Another method employed was the following: Across a sterile and well dried agar plate three parallel streaks were made with a cotton swab moistened with sensitive culture. The streaks were so placed that when a growth of 12 to 18 hours had occurred the approximating edges would be from 1 to 3 or 1 mm. distant from each other. Then at once, in the middle of the central streak (which should be about 8 mm. wide, as first laid on), a minute drop of the lytic filtrate was planted, so placed that the lytic action would extend through the central streak toward its free edges. If the principle was then to spread further, having reached the edges of the culture of the middle streak, and was to reach the first and third streaks, it would need to pass over the intervening stretch of clear agar. The results of many such tests, often modified in one way or another were the same as those mentioned previously for the colony experiments. The lytic agent readily passed over such areas of clear agar if they were not broader than 2 to 3 mm., and then attacked the fresh culture of the first and third streaks, or sometimes of one streak only, if the other interval was too great. From such freshly invaded areas the principle could always be obtained whenever the culture appearance indicated lysis or was touched by the a zone; also in a few cases in which the fresh culture appeared quite normal.

In observing these cases it was a point of special interest that the rate of progress of the bacteriophage across open agar spaces was essentially the same as the rate of progress through the culture mass itself. And this statement refers not only to the progress as indicated by the area of complete lysis and of secondary colony formation, but to the progress indicated by the gray zone * which, under certain conditions, surrounds the central area of lysis. This statement also holds for the translucent β zone which may often be observed surrounding the a zone. Thus, if the original drop-inoculation with the bacteriophage was a perfect circle, the gradual development of the lysed area, as well as of the a and β zones, gave a perfectly symmetrical picture, regardless

---

* Further consideration of the zonal phenomena (a and β zones) may be found later in this section.

of the passage of the bacteriophage over agar or through culture. This result could naturally be brought about only if the bacteriophage migrated with equal speed over or through the respective mediums. The conditions that might modify the appearance of this phenomenon have not yet been studied, and the actual mechanism of the passage remains undetermined.

Lysis under the Microscope.—One might anticipate that following the course of lysis microscopically would reveal some phenomena of significance. D'Herelle and Jeantet were the first to make observations of this sort and to report (d'Herelle) on certain granulations appearing in the cells under the dark field. These granulations, in turn, appeared to dissolve in the medium, undergoing, as d'Herelle believed, final lysis. The latest conclusion of d'Herelle is that none of these granular bodies probably represents the actual bacteriophage. It was noted that the cells swelled previous to lysis, thus producing globoid forms which finally ruptured. D'Herelle regarded the "bursting" process as evidence that the force operating was internal. Some of the staphylococci undergoing rupture had a diameter two or three times that of normal cocci.

At a later date the dissolution of cells was noted by da Costa Cruz, Pondmann, Hauduroy and Wollman. Contrary to the report of d'Herelle, da Costa Cruz stated that the cells appeared normal in shape and size up to the moment of lysis, and that the phenomenon revealed nothing of special interest. The cells about to undergo lysis could not be differentiated from normal cells. After lysis he saw only a slight residue. He concluded that this phenomenon probably deals with the permeability of the cell membrane and is of a purely physical nature. Wollman also followed the behavior of lysing cells and their appearance in fresh and in stained preparations. He also described the generation of enormous bacterial forms, some attaining a size even up to 15 microns long and 4 to 5 microns broad. In lysis, the form of the organism disappeared and was replaced by a mass of granules clearly marking the site of lysis. In B. coli Wollman did not observe the hypertrophy characteristic of the Shiga bacillus, but this feature has been clearly described and pictured by Larkum. Wollman called special attention to the circumstance that all the hypertrophied cells were cells in the course of active division. He also pointed out that the lytic reaction is related in a significant manner to important changes in the staining reaction of the cells. Similar changes were observed by Preisz. Béguet, also Bronfenbrenner, noted hypertrophied cells accompanying the lytic

Conclusions.—From the data presented in this section it would seem that practically no advance has been made in our knowledge, beyond the point of d'Herelle's observations published in 1926, so far as the physical characteristics of the bacteriophage are concerned, although certain important observations of Kuhn [222] will be mentioned later. I believe that it may be accepted that a definite bacteriophagic unit or corpuscle exists, but regarding its size and form little is known, although its dimensions probably lie between 30 and 90$\mu\mu$. It seems, moreover, knowledge on these points is likely to be lacking until adequate and reliable methods for the measurement of ultramicroscopic particles are devised. The evidence from filtration experiments is at present of slight value, except in a comparative way, for there are too many variable factors. Some of these are already recognized, but the difficulties and uncertainties occasioned by them are still unsurmounted. Still other factors may not yet have been recognized. Plasticity and adsorption on other colloids may be important factors. Moreover, one might inquire regarding the possible influence of motility of particles on filtrability. Does the bacteriophage unit possess motility? We do not know. It seems highly improbable, but the question possesses some significance if the bacteriophage is regarded as a living thing.

Final conclusions are also impossible regarding the nature of the electrical charge that might be carried by the lytic units and the possible significance of such charges in the binding of the lytic units to the cellular substratum. The weight of evidence is probably in favor of a negative charge, but the conclusion is by no means assured. In view of the experiments of Meissner, one may well postpone all conclusions regarding the manner in which the charge may affect the binding of the lytic units to the cell; or, indeed, whether the nature of the charge affects the binding at all. Although several writers have made use of a terminology implying an analogy between the binding of the bacteriophage to the cell and the binding of agglutinin (Meissner), there is little evidence to support such a view. It is not known that "the first stage of lysis" is in any way a phenomenon of absorption. I believe that errors may be made if one attempts to interpret what happens in bacteriophagic reactions in the light of the supposed knowledge of the mechanics of serologic reactions.

Regarding the manner in which we may picture the moment of lysis, the few available data suggest that the force exerted to break down the cell comes not from without but from within. Evidence which I can support from personal observations under the dark field indicates that in the few minutes preceding actual lysis there are to be seen in the Shiga cultures unusual numbers of large, swollen cells such as those depicted by d'Herelle and shown more clearly by Wollman, by Kuhn and by Larkum. In this connection, however, it is desirable to point out that this type of cell is not apparently different from some of those giant forms that have been reported by many observers in cultures manifestly undergoing dissociative reactions; nor is it apparently different from cells that can be observed, in small numbers, in cultures which manifest normal stability. It is also, perhaps, not without some significance that a sort of reaction which might be characterized as an "endo-sporulation" of enlarged cells and zygospore-like bodies, sometimes accompanied by the liberation of minute motile forms, has been described by Hort,[352] Kuhn,[221, 222] Mellon,[352] Enderlein,[117] Koch and Ziegenspeck and Koch for apparently normal, sensitive cultures or cultures in the process of dissociation.

4. THE RANGE OF ACTIVITY OF THE BACTERIOPHAGE AND SOME OF ITS CHARACTERISTIC MODES OF ACTION

Since the discovery of the Shiga bacteriophage by d'Herelle, similar agents have been found for an increasing number of bacterial species, although none of the obligate anaerobes or spirochetes is numbered in the list. Both gram-positive and gram-negative organisms may be affected. In some other species (Bacillus anthracis, Monilia) bacterio-phage-like reactions have been observed, but d'Herelle, as also the authors themselves, have disclaimed any relation to the phenomenon of transmissible autolysis because of the apparent absence of transmissi-bility. Some of these cases will be examined later in greater detail, because they concern a point of much interest relating to the possible existence of different modes of expression of the lytic reaction in different bacterial species. In addition, there have come to light other curi-ous autolytic reactions in cultures, living or dead, some of which manifest one or another of the characteristics of bacteriophage action, but all of which lack some of the important features. These reactions possess an independent interest, and some must be carefully differ-entiated from transmissible autolysis.

The range of reaction of a single strain of bacteriophage, whether monovalent, bivalent or polyvalent, is also a question of much interest and one that must be taken into consideration in any discussion as to the nature of the lytic agent. As will be seen, this phase of the problem is developing important aspects in relation to the range of serologic affiliations manifested by the O and R forms, as opposed to the S type, of various bacterial species, and its parallel in the bacteriophage reactions. These points, together with certain peculiarities of the lytic reaction, will be considered in the present section.

REFERENCES

Range of Activity: D'Herelle;[109] also for the following species: Sierakowsky and Zajdel (B. acidophilus);[312] Pico;[281] Pesch;[273] * Katzu;[209] * Brown and Basaca (B. anthracis);[60] * Sonnenschein (Monilia);[314] * Coons and Kotilia (B. carotovorus);[92] Israelsky (B. tumefaciens);[200] Hadley and Dabney (B. lacticus);[160] a Koser (a thermophilic organism);[216] Elder and Tanner (a psychrophilic organism);[144] Broudin (Pasteurella avium);[331] Clark and Clark (S. mucosus);[89] Schwartzmann (streptococcus from erysipelas);[308] Jeney (B. cloacae).[360]

Relation to Other Forms of Lysis: Gratia and Dath[158, 159] (by molds); Gratia and Dath (by Streptothrix);[160] Gratia and Rhodes[161] a (lysis of dead cells); Gratia[157] (by a strain of B. coli); Jaumain[201] (of staphylococci in sealed tubes); Rosenthal[297] (by Tyrothrix); d'Herelle[190] (of cholera vibrios by substances in intestinal fluids); Bruynoghe and Dubois[70] (by a typhoid strain); Ørskov and Larsen[147] (colony lysis in a paradysentery culture); Fleming;[130] Fleming and Allison[130] (the "lysozyme"); Reynals[298] (of dead staphylococci in sealed tubes by living organisms); Vignati[325] (B. coli inhibited by young culture of B. typhosus).

Size of Plaques: Bail;[26] Bruynoghe and Wagemans;[73] Bail and Watanabe;[27] Watanabe;[328] Gratia;[155, 156] Gratia and de Kruif;[141] Hadley;[165] Asheshov;[17, 17] a Bronfenbrenner and Korb;[54] d'Herelle;[193] Kline;[363] Hadley and Dabney.[199] a

Individual Characteristics of the Bacteriophage: Wollman and Wollman (retains susceptibility or resistance to trypsin after adaptation to new host); Burnet (biologic independence of certain strains).

DISCUSSION

Range of Reaction among Bacterial Species; Bacterial Convergence. —D'Herelle[325] in his latest book presented a list of 24 or more organisms for which a bacteriophage had been discovered up to 1925. To this list may now be added lytic agents for B. acidophilus, B. carotovorus, B. cloacae, B. tumefaciens, B. lacticus, the thermophilic bacillus of Koser, the psychrophilic bacillus of Elder and Tanner, Pasteurella avium and S. mucosus. Whether the bacteriophage reported by Dutton[110, 111] can properly be included in this list is not clear. There should also be added the lytic agent for B. pyocyaneus (Canzik[84]) which d'Herelle has not accepted as entitled to recognition, but which I have shown[224] must be regarded as a form of the bacteriophage.

Regarding the range of activity of a single, pure-line bacteriophage, little can be added to the data already presented by d'Herelle. Each

---

* Authors do not admit the presence of the bacteriophage in the cultures described.

strain seems to be able to influence a small number of organisms that belong to the same group. In this connection, d'Herelle reported that the Shiga bacteriophage is alleged to have been adapted to act on unrelated organisms, such as M. aureus, and there are a few other reports of this nature. These cases will be considered later, since they demand re-examination from a different point of view.

What determines the range or the limits of action of a single strain of bacteriophage is a problem of much interest, because it may have a definite bearing on the nature of the lytic unit. Up to the present there has not been any interpretation other than that which d'Herelle has related to his virus theory.

The studies of Schutze [302] in 1922 laid the basis for a new understanding of many curious heterologous serum reactions, in which he referred to the phenomenon by the term, "serological cosmopolitanism," or bacterial convergence. The later work of Bruce-White,[61] of Goyle [145] and of Balteanu [352] clearly indicated that these heterologous reactions were mainly due to the presence of the O or the R antigens in the cultures concerned. Since that time other investigators have shown that the often-observed community of antigenic structure is mainly referable to the heat-stable antigens (Felix [352] and Felix and Olitzky [352] on qualitative receptor analysis).

The interesting question has, therefore arisen regarding the possible parallelism between bacteriophagic and serologic reactions shown by bacteria. Could it be demonstrated that bacteria of different species but related serologically, revealed at the same time a similar bacteriophagic relationship? I first studied this point in three species already known to possess somewhat curious serologic affiliations; namely, B. typhosus, Bact. pullorum and B. gallinarum, the last two being paratyphoid cultures pathogenic for birds. A bacteriophage was produced separately for each species and tested against the other two. It was clearly shown that the bacteriophagic relationships paralleled almost exactly the course of the serologic reactions, thus seeming to imply that the community of bacteriophagic and serologic reactions was dependent on a community of antigenic structure. In these experiments it should be added that the smooth S type cultures were employed, and these, as is known, reveal a slighter degree of heterogeneity than the reciprocal O and R forms.

In 1926, also, Marcuse [245] closely approached the same problem, though in a different manner, in a study of the serologic and bacteriophagic affiliations of a certain strain of B. dysenteriae Y. From this culture was obtained a strong bacteriophage which lysed certain coliform strains as well as itself. It was also noted that these sensitive colon strains had the power to absorb the agglutinins from a serum prepared against the original Y strain. Coliform cultures which were resistant to lysis did not absorb these agglutinins, or absorbed to a smaller degree.

In 1927 F. M. Burnet [78] recognized the importance of the problem of the possible parallelism of the bacteriophagic and serologic reactions and performed experiments much more extensive than my earlier ones, and experiments which had the merit of taking into consideration the R type antigens. It seems to me that these tests of Burnet establish the underlying principle of parallelism, at least for the members of the typhoid-paratyphoid group; and what happens here can be regarded as possessing a general significance, although Burnet was cautious in his conclusions. Burnet worked with strains of B. typhosus, B. enteritidis and Bact. pullorum. He employed both the S and the R forms of the cultures and concluded that the bacteriophagic reactions of these three species were practically identical. From a consideration of their sensitivity to a series of bacteriophage strains, they formed a homogeneous group. But the important point for present purposes is that they also showed similar serologic reactions, manifestly determined by the presence of the common, heat-stable O or R antigens. Burnet found that strains of bacteriophage developed on rough enteritidis cultures were active against rough variants of most of the Salmonella types studied, the range corresponding to the parallel extent of the common, heatstable agglutinogens among these forms. Strains without O agglutinogens in common did not show sensitivity to the enteritidis bacteriophages. Burnet thus saw a coordination between agglutinin absorbing capacity and the capacity of an organism to absorb bacteriophage, and believed that, on the whole, this pointed toward a "biological independence" of the bacteriophage particles, in contrast to biologic unity, as maintained by d'Herelle.

The conception of the unity of the bacteriophage is further opposed by Koser's [216, 217] discovery of a bacteriophage for a thermophilic species. In this case the agent operates at a temperature that is destructive to the majority of the nonspore-forming bacteria. It is also opposed by the discovery by Elder and Tanner [114] of their psychrophilic bacteriophage which is active at temperatures around 4 C. (39.2 F.). Such observations cannot fail to convey the suggestion that the specific bacteriophage is closely related biologically to the culture from which it arises, partaking equally of its most characteristic growth features.

This offers the suggestion, already supported by other evidence, that the protoplasm of the lytic corpuscle is in some way continuous with that of the bacterial cell. If we could regard the effective union of the bacteriophage with the sensitive cells in the light of a process of fecundation, occurring readily within the same species but also to a limited extent between species less closely related, many aspects of the problem would become clearer. It is possible that we shall yet come to the view that there exists in bacteria a reproductive mechanism making possible hybridization of a sort; this phenomenon has been treated by the Swedish bacteriologist, Almquist,[3] and supported by many of the observations of Enderlein [116, 117] in connection with bacterial cyclogeny.

At any rate, the observations on the relation between bacteriophagic and serologic reactions present for the first time a tangible biologic suggestion why a certain bacteriophage is able to influence some species but not others, and does not leave the matter entirely to the gastronomic vagaries of a hypothetic bacteriophagic ultravirus, looking about for the antigen of its predilection. It should be added, however, that this conception does not explain why, as is sometimes the case, a bacteriophage may be active for certain strains of a species, but not for other strains of the same species.

Variation in Size of Plaques.—According to d'Herelle's earlier views, the size of plaques is determined by a number of conditions, many of which seem to be identical with those determining the size of bacterial colonies. This involves such points as the condition of the medium, its thickness, the density of the bacterial culture, the crowding of the lytic corpuscles, etc. But for d'Herelle the size of the areas was chiefly correlated with the "virulence" of the bacteriophage.

That the problem of the size of plaques was not so simple as this, however, was first indicated by the observations of Bail, and of Bail and Watanabe, on certain differences in their so-called "elementary strains" of the lytic agent. These investigators noted the existence of three sorts of plaques, the large, the medium and the small, which differed in their action on sensitive cultures and also, as shown later (1923) by Watanabe alone, in the serologic reactions of the bacteriophage contained. The serologic aspects will be considered later; and, for the moment, I will turn to the extension of these observations made by Gratia in 1923. I shall consider some of these points in detail because their importance has not been emphasized in more recent publications dealing with the size of the lytic areas; moreover, because they have not received any appreciable consideration from d'Herelle.

Gratia and de Kruif in 1923 showed that subculturing from areas of different size on coli plates yielded lytic filtrates of very different lytic activity. Some pure-line strains were "weak" while others were "strong." The "weak" filtrate conserved its properties, while the "strong" filtrate, even after four successive isolations, and like the original lytic agent, divided into a stable (weak) and an unstable (strong) principle. The weak principle was characterized by the production of small lytic areas, while the strong principle was represented by large, or large and small, areas. An interesting question thus arises regarding the actual differences between these two types of the bacteriophage which are active for a certain strain, and which I have found equally in the case of B. dysenteriae Shiga, B. typhosus, B. paratyphosus A and B and several other species; which also my student, Cameron, has demonstrated for B. proteus X19. To obtain light on these matters inquiry must be diverted to the consideration of another subject.

In 1921, Arkwright [8] had shown the existence in "normal cultures" of two culture types, the smooth (S) and the rough (R), the former causing a homogeneous clouding in broth, the latter an agglutinative form of growth. In 1922, Bordet and Ciuca [46] showed that the R type culture (Bordet's "P") was more resistant than the S type (Bordet's "B") to their weak lytic principle; and the same was shown in 1924 by Arkwright,[9] who also pointed out that a number of variants, some of which were sensitive, could be regained from completely resistant strains that had arisen without the intervention of the bacteriophage. We shall now see that these two distinct forms of culture (cyclostages in the developmental history of the species), S and R, are related in a peculiar way to the phenomena of the plaques.

Following the contribution of Gratia and de Kruif, Gratia [155] extended his original observations in a report dealing with the so-called "heterogeneity of the coli lytic principle." He called attention to the existence of three types of coli bacteriophage, differentiated according to the size of the plaques and the energy of lytic action. These were:

Principle giving only small areas (weak).
Principle giving only large areas (relatively weak).
Principle giving large and small areas (strong).

When a drop of the "large area" principle was placed on agar, along with the sensitive (S) culture, the track of clarification revealed many resistant secondary colonies which were large, flat, rough, nonfluorescent and, in broth culture, self-agglutinative. If, in a similar manner, the small area principle was placed on agar with the S culture, the track of clarification became well covered with secondary colonies; but these were now small, convex and glistening; and, when seeded into broth, gave a diffuse form of growth like the S culture. It thus appeared that the nature of the secondary, resistant culture varied with the sort of bacteriophage employed. Gratia, however, observed another important fact: The colonies that were resistant to the large area principle were sensitive to the small area principle; and, conversely, the culture resistant to the small area principle was sensitive to the large. This observation was sufficient to explain why a mixture of the two principles constituted a "strong" lytic agent. The resistants to the one were lysed reciprocally by the principle of the other. It is also of interest that Gratia observed that the two principles possessed different points of heat inactivation. The large area principle was destroyed at from 60 to 62 C., while the small area principle resisted a temperature of 70 C. These two forms of the bacteriophage also possessed interesting serologic differences, as had also been observed by Watanabe. This aspect of the subject will be considered later.

In completing this phase of the study of Gratia, it may be added that he also succeeded in isolating from "normal" coli cultures an R form which not only possessed all of the external characteristics of the resistant, secondary culture left by the lytic agent of the large area type, but which, like it, was also refractory to the large area principle. At the same time it was sensitive to the small area principle. In addition, Gratia isolated from a "normal" coli culture a diffuse strain (smooth S) which, contrary to the agglutinating type (rough R), was sensitive to the large area principle and refractory to the principle of the small areas. He thus demonstrated the existence, in "normal" culture, of two strains characterized by different grades of resistance; but he also pointed out that there were some strains that were equally sensitive to the two forms of bacteriophage, this being true of a large number of "diffuse" cultures. Only such strains were able to regenerate the principle "in its entirety." It thus sufficed, in order to obtain a small area principle in pure state, to regenerate the original bacteriophage at the expense of a self-agglutinating strain of coli; or, in order to obtain a large area principle, to regenerate the original bacteriophage at the expense of the "normal," diffuse type of culture.

These results led Gratia to the view that the notion of "weak" or "strong" principle is purely relative and depends on the strain of bacteriophage employed and on the type of culture submitted to its action. To place the matter more or less in his own words, a large area principle will thus appear inactive and feeble, or active and strong, according to the type of organism submitted to its action; whether it is constituted mainly of resistant, agglutinative germs or contains none of them. The author continues these views with the surmise that B. coli is manifestly susceptible to many other variations than those of the "diffuse" (S) and the "agglutinative" (R); and for these other variants it is possible that other variations of the lytic agent may exist. If this should be verified, he continues, then, to the veritable "mosaic" of culture types which the coli culture represents, there will be found a corresponding "negative mosaic" of bacteriophage. Thus, in a general way, for every heterogeneous bacterial species there will be a corresponding heterogeneous bacteriophage.

In 1924, having overlooked Gratia's important studies, I [165] reported briefly the differentiation of lytic areas of Shiga into the large and the small, without the apparent existence of areas of intermediate size. The small area principle "bred true" on continuous passage, but the large area principle, in time, invariably gave rise to both large and small area strains. In 1925, Bronfenbrenner and Korb,[54] apparently overlooking Gratia's contributions and all earlier work, as well as my own, considered the problem of the size and number of plaques, as determined by the condition of the medium. They discussed the variations, but it is apparent that these writers unfortunately dealt exclusively with a strain of bacteriophage of the small area type. They thus did not touch on the question of the large versus the small plaques. One may gather from their results, however, that, even within the small area type, there may exist some slight, though insignificant, range in the size of the plaques.

Quite recently, Kline presented data (in connection with a study of the spontaneous origin of the bacteriophage) on the occurrence of large and small plaques in streaked cultures of B. typhosus and other intestinal bacteria. In general, the large areas were about 3.5 mm. in diameter, and the small 1 mm. or less. There were not any plaques (within the same pure strain of bacteriophage) which were intermediate in size; in other words, the variations were discontinuous. As a rule, the large areas were found to give rise to large areas only, and the small areas to small only. The author seemed to doubt the origin of the small plaques from pure large area strains of bacteriophage, as first demonstrated by Gratia and later confirmed by myself without the knowledge of Gratia's work; but the matter was not thoroughly tested by Kline. He is right, however, in concluding that his results "seemed to indicate the existence of two distinct lytic mechanisms . . .," and he was able to demonstrate their mutual independence. These "mechanisms" he termed "B" (for large areas) and "b" (for small areas). Kline believed that his results might explain the observations of Seiffert to the effect that the acquisition of resistance to one lytic agent might be accompanied by acquisition of sensitiveness to another. Kline also regarded his results as confirming the conclusions of F. M. Burnet, that, in the lytic process, more than one "lytic ferment" is operative. The author was manifestly unaware of the earlier work of Gratia, and did not correlate the behavior of his two "mechanisms" with the S and R forms of culture. His results and conclusions, nevertheless, furnish an interesting confirmation of several important and earlier observations. (Further consideration of the two different principles [α and β], producing the large and the small plaques, respectively, may be found in section 8.)

Since 1924 certain studies on variations in size of lytic areas have been in progress in this laboratory. With my student, Eugenia Dabney,[169a] I have already reported on certain features concerning the dual nature of an antiparatyphosus B bacteriophage. Here there were obtained areas of two sizes: the large, having an average diameter of 6 to 7 mm., and the small, having an average diameter of 0.5 to 2 mm. Only a few areas of intermediate size appeared; and these, by their form and behavior on subsequent propagation, manifestly belonged to the large area type. In other words, the variations were discontinuous and there were no true intermediates. Frequently, under conditions of crowding on the agar plates, the large areas might enclose one or more of the small areas. In such cases the small areas were especially distinct if they occupied a position in the outer zone of the large areas. Here, the small areas manifestly completed the lysis that had been left incomplete by the action of the large areas. In order to simplify the terminology dealing with these two principles and their respective areas, we have termed the principle determining the large areas as the alpha (α) and the agent determining the small areas as the beta (β) principle. Similarly the large and small areas have been termed the α and the β areas, respectively.

By selecting discrete areas of the α and the β type, pureline α and β principles, respectively, were built up. When the α principle was tested against the original S type culture, only large areas were produced; and, when the β principle was tested against the same culture, only small areas were produced. After continued propagation, however, while the β principle continued to "breed true" indefinitely, the α principle, after a time, began to produce the β areas; that is, to generate the β principle in addition to the α principle. With continued propagation by serial feeding and filtration, it often happened that an α principle became nearly or wholly transformed into a β principle. We also observed that the α principle was active against cultures which arose (after lysis) as secondary resistants to the β principle; moreover, that the β principle was active on cultures that arose as secondary resistants to the α principle. The action of the two principles was thus "reciprocal," as Gratia [156] had earlier shown for his two components of an anti-colon bacteriophage. In certain cross-tests in which the anti-paratyphosus α and β principles were employed against an S type typhoid culture it was only the β principle that registered, thus yielding only the small areas; the typhoid culture was not affected by the paratyphoid α principle. Moreover, when the α and β paratyphoid principles were mixed in vitro and then applied to the typhoid substratum, only β areas appeared on the plates. At the same time, when a β typhoid principle was tested against the S type paratyphoid substratum, only small areas appeared. This typhoid β principle was also active against the secondary resistants (paratyphoid) arising from the action of the α paratyphoid principle on S type paratyphoid culture, but not against secondary resistants (paratyphoid) arising from the β paratyphoid principle acting on a similar paratyphoid culture.

A point of considerable interest lay in the difference in the points of thermal inactivation of the α and β paratyphoid principles. When heated in the waterbath in long, drawnout capillary tubes (with the ends sealed) it was ascertained that the α principle was inactivated by heating for 30 minutes at 63 C., but not at 60 C.; that the β principle was rendered inert by heating for 30 min. at 75 C. but not at 70 C. This difference might seem to possess special significance in view of the fact that the heat-labile α principle is generated mainly at the expense of the culture type (S) which is characterized by the possession of a dominant heat-labile antigen, while the heat-stable β principle is generated mainly by the culture type (R) which is characterized by the possession of a dominant heat-stable antigen. In these cases the temperature which serves as the criterion of heat-stability or heat-lability of the antigens is 100 C. (boiling) for one-half to one hour.

That the observations presented above are not exceptional in bacteriophage reactions is evinced by the fact that many of the points mentioned for B. paratyphosus B have been verified and extended by my student Bonaventura Jiménez with reference to the α and β fractions of a bacteriophage active on B. paratyphosus A. At the same time it must be added that several aspects of the problem dealing with the relation of the α and β principles to the various cyclostages of the homologous culture are still far from clear and demand much further study. In all this field, however, it is an uncontrovertible fact that the finer details of the reactions lie within the realm of the dissociative aspects of bacterial behavior.

Zonal Characteristics of the Large Areas.—One interesting feature differentiating the large and small plaques formed in a sensitive culture substratum on agar plates involves the following point which I have studied at intervals since 1923. A "mixed" lytic agent for B. paratyphosus A or B was obtained by the use of sewage filtrates. The substratum culture was naturally of the S type but presumably contained some S and some R organisms, although the appearance remained smooth. The first isolation filtrate obtained, when properly diluted, impressed itself on the substratum (on plates or slants) by the formation of large areas mainly, and these attained a diameter of from 10 to 15 mm. With subsequent feedings and filtrations the second (small) type of plaque made its appearance and the "small" units seemed to increase with each succeeding filtration. These small areas in the same preparation usually revealed little variation among themselves. The striking contrast in size of plaque was greatest in the paratyphoids, less so in Shiga, typhoid and coli. As a rule, there were not any intermediates in size. The difference, however, which I wish particularly to mention at this time relates, not to the size of the areas, which has often been noted, but to a peculiar zonal phenomenon appearing around the plaques. In the small plaque formed on S culture the condition is the simplest. Here one observes merely a small bare area around which a faint gray zone, perceptible only under the hand lens, may exist. Outside this zone is the fresh and apparently unmodified culture. In the large plaques the appearance is quite different. In the center is the bare spot (sometimes containing a sprinkling of secondary colonies), but this is immediately surrounded by a zone of distinctly modified culture. This zone is grayish and more translucent than the fresh culture. The line of division between this zone and the bare area is somewhat indefinite, while the line of division from the outside sensitive culture is fairly distinct. The width of this gray zone varies from 1 to 3 or 4 mm. I have termed this the α zone. But, outside this, there may sometimes appear a second zone which I have termed the β zone, and which is observed only surrounding the large plaques. It can be seen to best advantage in artificial lytic colonies produced by point inoculations of the lytic filtrate in the center of an agar plate previously coated with sensitive culture. This is much less distinct than the α zone. It begins as a faint bluish expansion of the α zone and extends itself into the fresh, sensitive culture. It progresses evenly on all sides of the central lytic spot and, after a week or two, may have a diameter almost equal to that of the plate itself. Under these conditions it is, of course, necessary to start the lytic colony in the center of the plate. The agar should be at least from 7 to 9 mm. thick to permit continuous development of the culture over a prolonged period. Thus, proceeding from the center of the lytic spot toward the rim of the plate, one encounters, in turn, lytic area, α zone, β zone, fresh culture and, finally, bare agar surface. The β zone, so far as I have observed is never present except around large plaques and in the smooth culture (S) substratum. It reveals itself to greatest advantage around artificial plaques (made by point or small drop inoculations) having a diameter of 2 cm. or more. It does not appear in substrata composed of R type culture.

A question of interest is the presence or absence of the bacteriophage in the α and β zones. I studied this point in 1925 with an homologous bacteriophage acting on a sensitive coli-like culture from river water. It appeared that the α zone always contained active lytic agent. Samples made at various distances into the β zone showed the penetration of the bacteriophage over a distance of 6.5 mm., but not 7.2 mm. In this case the β zone had a breadth of 22 mm., and from other points in this area only normal sensitive culture was isolated. The positive tests for bacteriophage in the β zone were obtained 12 days after establishing the lytic spot.*

The Relation of the α and β Principles to Zonal Action.—Although some of the features of the α and β zones have been considered on a previous page it is desirable to mention this subject again with relation to the mode of action of these different principles on the S and R forms of culture separately.

First it should be pointed out that the studies conducted especially by my students, Armstrong and Jiménez, have shown that, in a para-

---

* These experiments and numerous others bearing on the size of the lytic areas and on the peculiar zonal phenomena have not been published earlier because I had hoped to obtain some information that would lead to an understanding of them. As years have passed, however, I have not been able to gain any understanding of the problem. For this reason, I anticipate presenting the results as they stand at an early date. I believe, however, that they concern the relation of the bacteriophage phenomenon to microbic dissociation, and in a significant way.

typhosus A culture at least, the α zone phenomenon is produced by the α principle only. Moreover, it may occur whether the substratum culture is of the S or R type.

The β paratyphoid principle, on the other hand, produces no α zone; and this is true whether the substratum is represented by the S or the R form of culture. In both cases the line of demarcation between the lysed and unlysed areas is sharp and clearcut, rather than indefinite and hazy, as that produced when the α principle is employed on the same cultures.

These results seem to show that the mechanism effective in the formation of the α zone is not primarily dependent on variations in the form of culture but on differences inherent in the respective principles themselves. I may add that this difference also holds for the α and β principles of B. paratyphosus B, the Shiga dysentery bacillus and probably other bacterial species.

One other somewhat related difference in the action of the two principles may be mentioned here. As already noted, the areas of lysed and unlysed culture, acted on by the α principle, are separated by a hazy, gray zone several millimeters in breadth. Behind this barrier, that is, in the normal culture substratum, isolated α lytic areas are seldom found. The areas of lysed and unlysed acted upon by the β principle, on the other hand, are separated by a clearcut line of cleavage showing apparently clear agar (or isolated, secondary colonies) on the one side and normal culture on the other. But, under these conditions, there are commonly observed, lying inside the border of apparently normal growth, scattered lytic spots of the β type, sometimes few and sometimes numerous. Such lytic areas may occur even four millimeters inside the border line where the chief area of lysis ends abruptly.

How is this further invasion of the culture substratum by the isolated β units effected? Is it the result of accidental spattering when the principle is laid over the agar surface? Is it due to a diffusion or flow of the culture liquid over the surface of the moist plate, carrying the lytic corpuscles into the interior culture regions? Do motile bacteria, still active for a time after "invasion" by the lytic units, carry these units back into the culture? Do some of the lytic corpuscles penetrate the interior culture regions by means of their own power of movement? If any of these explanations hold for the β principle, why are the same phenomena not observed in the action of the α principle? Are some units endowed with power of motility while others are not so endowed? No answers are at present available for any of these questions.

Relation of Bacteriophage Action to Other Forms of Lysis.—The study of the bacteriophage has stimulated a renewed interest in bacteriolytic phenomena in general, and many new instances of the disappearance of bacteria through lysis have been reported. Some of these have dealt with the lysis of living bacteria, others with the lysis of dead bacteria. Some have concerned the effects of bacterial associations, and a few have dealt with the influence of living cells on dead bacteria; while, in the majority of cases, the bacteriophage, at least in the usual sense, has apparently been absent, and has been so reported by the investigators themselves. There are some cases, however, in which, although its presence has been denied, its influence may be suspected.

Gratia and Rhodes reported the lysis of dead staphylococci by living organisms, and Jaumain has showed the lysis of living staphylococci in sealed tubes. Bronfenbrenner discussed the lysis of dead bacteria. Gratia and Dath demonstrated the lysis of bacteria by molds (Streptothrix). Gratia described a curious strain of B. coli whose filtrates caused the lysis of another strain; and a similar instance was reported by Bruynoghe and Dubois. Mention should also be made of the interesting and well known studies of Fleming and Fleming and Allison on the "lysozyme," which shows certain similarities to the bacteriophage but which differs in important details from the classic mode of action; also the lysing paradysentery culture of Ørskov and Larsen, and the lysis of dead staphylococcus reported by Reynals. Further, Vignati, in 1926, reported the inhibiting action of a young culture of B. typhosus on B. coli, interpreted as a microbic antagonism. Rosenthal reported that a culture of Tyrothrix was able to dissolve a great number of bacterial cultures, both living and dead. D'Herelle has called attention to a lytic substance, not bacteriophage, in the intestines of patients with cholera, which is active on the vibrio.

The exact nature of these reactions is still unknown. In some, merely digestion of the cells by enzymes may be concerned. In others, there is evidence of dissociative action. Many of them d'Herelle relegates to the category of "bacterioclysis," a "disease of bacteria." Here also he places the lysis of B. pyocyaneus and of B. anthracis. Taking these instances as a whole, however, it may appear that the so-called lysis of living bacterial cells is a phenomenon embracing several distinct reactions, some of which may be actually disintegrative, while others are undoubtedly concerned with cell transformations at present little known or understood. The latter aspect of the subject is considered further, on a later page.

Conclusions.—From the data presented in this section the following conclusions may be derived. Regarding the so-called adaptation of a strain of bacteriophage to a widely separated species, as d'Herelle's reported success in adapting a Shiga bacteriophage to M. aureus, and other similar instances, I believe that all these successes must be regarded with doubt; that such adaptations have occurred is highly improbable. When they have seemed to occur, it is more probable that a lytic agent for the new species would have been obtained in any case, and without the addition of any heterogeneous bacteriophage filtrate, if the same number of serial filtrations or transmissions had been performed. The evidence supporting this view was presented in an earlier section, but it may be said in passing that many experiments, such as those performed by myself with Klimek and Kiesewetter, and by others, clearly demonstrate that all that is necessary, often, to generate bacteriophage in normal, sensitive culture, is to grow the culture repeatedly in its own filtrates. Naturally, this possibility detracts considerably from d'Herelle's conception of the "unicity" of the lytic agent, and the consequent necessity of adaptations to new hosts. That something called "adaptation" does occur, however, within a limited group of closely related organisms, cannot be doubted. Regarding the possible basis for this range of reaction, it appears that F. M. Burnet has made a contribution of special significance in establishing, within one bacterial group at least, the restriction of range of lytic action of a single strain of bacteriophage to those bacterial species endowed with the common O or R antigens. It can scarcely be doubted that this work will soon be confirmed and extended into other bacterial groups.

The problem of the size of the plaques, and their relation to the smooth (S) and rough (R) forms of culture, is one the significance of which has been to a large measure overlooked by nearly all investigators. The results of the work of Bail, Watanabe, Gratia, Asheshov and Kline, although clearly bringing out certain facts, at the same time render the nature of the diverse type of plaque even more complex than would appear from the simple statements of d'Herelle, Bronfenbrenner and others. That differences existing in the mediums are of much importance, as both these writers have amply shown, cannot be doubted; but the question is perhaps pertinent regarding the extent to which the condition of the medium modifies the aspect of the plaques by modifying the nature of the culture substratum. The effect of moist or dry agar on dissociative reactions is well known.

On the other hand, one must not lose sight of the fact that present evidence is sufficient to indicate that the lytic corpuscles differ in themselves. Bail believed that he could detect three distinct, "elementary" strains, characterized respectively by the production of large, small and medium sized plaques and by other differences in their mode of action. The work of Gratia, of Asheshov and of Kline and my own observations have shown, however, that there are, in all probability, but two elementary principles, those producing, respectively, the large and the small areas, while the strain producing the areas of intermediate size should be grouped with the large area strain. D'Herelle [193] gives only the most superficial consideration to these aspects of bacteriophage action, and especially with reference to Bail's and Gratia's fundamental contributions. His explanation of the observed variations in plaque size is based entirely on the inadequate notion of differences in the "virulence" of the bacteriophages concerned. Thus, if plaques of different size are observed, it is because the bacteriophages are endowed with "multiple virulencies." The most "virulent" strains are associated with plaques of the largest size, and the least virulent strains with plaques of the smallest size, assuming, naturally, a continuous series of variations between the two. This is a perfectly natural and inviting hypothesis; but, unfortunately, it is not supported by the facts in the case. Nowhere else, I believe, has d'Herelle championed so valiantly a conclusion which had little to support it but its logical probability. The facts which oppose themselves to d'Herelle's view are the following: 1. The large area principle is not necessarily a "strong" (virulent) principle. 2. The small area principle is only relatively "weak." 3. A really "strong" principle is one which is composed of both "large" and "small" units, and Gratia has revealed the reason for this. Moreover, the energy of action of either a pure "large" or a pure "small" principle depends entirely on the type of culture substratum (cyclostage) on which the principle acts. Thus, if the bacteriophage is a filtrable virus in the d'Herelle sense, it is a virus possessing a dual personality—an aspect of the subject which naturally still leaves much to be explained. As a matter of fact, none of the data at present available regarding these two different lytic units, representing the bacteriophage in its entirety, is interpretable on the assumption of variations in virulence of a parasitic, filtrable virus.

In an earlier publication I have expressed the opinion, and presented some evidence in its favor, that the whole problem of the bacteriophage is interwoven with the phenomenon of microbic dissociation. Further proof of this is found in the observations reported in this section   It is Gratia's work in particular, however, that demonstrates the striking interrelation between the large and small areas, the two varieties of lytic agent, and the smooth (S) and rough (R) types of culture. Thus do the two major cyclostages resulting from the dissociative reaction, namely, the S and the R cultures, together with their secondary resistants, determine clearcut and far-reaching differences in the attributes of the bacteriophage which they regenerate.

The virus theory of d'Herelle is no more able than the vitiation theory of Bordet and Ciuca to throw light on the nature or meaning of these peculiar reactions; or to explain their curious relation to important aspects of the dissociative phenomenon—even the existence of which (at least so far as their theories of the bacteriophage are concerned) seems to have escaped both of them. Indeed, while the virus theory can explain creditably many of the simple reactions of transmissible autolysis, it becomes each year increasingly unable to cope with the more complex reactions which have been shown to exist as important and intrinsic phases of the complete phenomenon.

5. THE INFLUENCE OF SPECIFIC SUBSTANCES AND OF CONDITIONS OF ENVIRONMENT ON BACTERIOPHAGE

In this section will be considered certain reports dealing with the influence of substances or environments, either on the bacteriophage directly or on the lytic reaction. The citations will be limited chiefly to those which aid in making clearer the nature of the agent, or to draw about it some possible limitations of the views. Little new knowledge has been gained regarding the influence of chemical substances or antiseptics. The influence of radiation, oxygen and temperature has, however, received further attention from investigators, and new facts have been derived. This is true also for the influence of commercial preparations of enzymes and of immune serums. As will be pointed out in the following pages, however, it has become necessary (in view of the knowledge of the dissociative reactions occurring in bacterial cultures under the influence of similar substances or conditions) to differentiate clearly between the influence of the factor concerned on the bacteriophage itself, and the influence of this factor on the organism, thereby often modifying the course of the lytic reaction.

REFERENCES

Alcohol: Hauduroy;[172] de Poorter and Maisin;[261] Watanabe;[328] [a] Bronfenbrenner and Korb;[53] Bronfenbrenner;[52] Callow.[83]

Antiseptics: Wolf and Jansen;[332] [a] Ordelt;[266] Callow.[83]

Bile: Joannides;[202] Hauduroy.[181]

Calcium Salts: Bordet.[37] Various Neutral Salts: Callow.[83]

Centrifugalizing: Hauduroy;[177] Wollman;[335] Applemans.[7]

Chloroform: d'Herelle;[192] Kabeshima;[203] Callow.[82]

Culture Type: Gratia;[136] F. M. Burnet;[78] Arkwright;[9] Eguchi;[119] Bordet and Ciuca.[45]

Distilled Water: da Costa Cruz.[94]

Enterokinase: Combiesco;[91] Wollman.[334]

Ether: Fejgin;[356] Stassano and Beaufort;[318] Nakashima; Callow.[83]

Electrolytes: da Costa Cruz;[97] Lisbonne and Carrère;[238] Ciuca;[87] Bronfenbrenner and Korb;[56] Brutsaert;[64] Zdansky;[366] Stassano and Beaufort [317] (action of sodium citrate).

Eau de Javelle: Arloing and Langeron.[10]

Fluoride: Bablet.[20]

Gelatin: Doerr;[106] d'Herelle;[190] Nakamura;[258] Brutsaert;[63] Hauduroy [178]

Light (ultraviolet): Applemans;[6] Biemond;[38] McKinley, Fischer and Holden;[253] Gildemeister;[137] [a] Gerretsen et al.;[136] Zoeller.[350]

Mercuric Chloride: Proca;[299] d'Herelle.[193]

Oxygen: Schwartzmann;[304, 305] d'Herelle.[194]

Pancreas: Keller;[212] Hoder and Suzuki.[196]

Pepsin: Wollman.[334]

Radium: Bruynoghe and Mund;[76] Lacassagne and Paulin.[226]

Reaction of Medium: d'Herelle;[186] Gratia;[302] Callow.[83]

Serum, Immune: See section on serology; also Gregorieff [162] Weiss.[330]

Sodium Citrate: Stassano;[317] Asheshov.[18]

Sodium Sulphate: Weiss [330] (precipitation)

Sugar Mediums: Weiss.[372]

Temperature: Koser;[214, 217] Elder and Tanner;[114] Gratia;[155, 156] Gregorieff;[162] Rivers;[296] d'Herelle;[193] Hadley and Dabney.[169] [a]

Trypsin: Wollman;[334] Arnold and Weiss;[16] Weiss;[330] Jötten;[214] Wollman and Wollman.[341]

DISCUSSION

Chemical Substances; Bile.—The work of Kabeshima, Hauduroy, d'Herelle and others on the influence of alcohol has been extended by Bronfenbrenner and Korb, by Bronfenbrenner and by Callow. D'Herelle had previously found that the lytic corpuscles were destroyed in less than 48 hours by 90% alcohol. DePoorter and Maisin had reported that even 70% destroyed the corpuscles in 24 hours. All the races studied by Watanabe were reported to have been destroyed by 75% alcohol in 30 minutes. Bronfenbrenner and Korb found that alcohol at room temperature destroyed lytic activity after from 6 to 24 hours' exposure. At 7 C. traces of activity remained after four weeks. This residual activity was transmissible, thus indicating that it was not due to an enzyme, as earlier proposed by d'Herelle. The reaction was regarded as analogous to the inactivation of certain enzymes and toxins. In 1926, the same authors showed that the addition of neutral salts to the lytic filtrates increased the rate of alcohol inactivation, while the removal of salts tended to render the lytic filtrate less sensitive. The authors believe that the inactivation rate by alcohol depends on the rate of precipitation of the coagulable constituents of the medium and is not due to direct toxic action on the bacteriophage. From these results the authors are led to a chemical view of the nature of the lytic principle.

Considering all the results to date, evidence seems to point to the fact that the lytic agent is precipitated by alcohol, along with protein matter in the filtrates, and that it may be completely destroyed or inactivated, depending on the concentration of the reagent and the time of exposure. In the absence of electrolytes, the effect of alcohol, as of acetone, is much reduced.

Regarding the independent action of electrolytes themselves, a considerable amount of evidence, and particularly that furnished by da Costa Cruz, implies that they are necessary for lysis. According to this author, distilled water flocculates the principle, but according to d'Herelle, this is observed only when the water is acid. According to Marcuse, even concentrated salt solution does not destroy the agent. Bordet studied the effect of calcium salts, and found that with one principle (L) some calcium (calcium chloride) was necessary for lysis. This seemed to be in agreement with the results of Stassano and Beaufort who showed that the hindering influence of sodium citrate on lysis could be neutralized by the addition of calcium salts. Asheshov, however, indicated that one could not generalize from the latter results, since some lytic agents are found which can operate in the absence of calcium ions, while others can be adapted to do so. This author showed, in addition, that small amounts of citrate might actually incite the lytic reaction.

Many other salts having some germicidal power have been found to destroy the lytic agent within a short time. Most recently Arloing and Langeron have reported that eau de Javelle, even in concentrations of 1 : 1,000, prevented the action of the bacteriophage for B. shigae, B. coli and B. typhosus. Since the bacteria grew well in the 1 : 1,000 dilutions, he believed that the influence was on the bacteriophage, rather than on the bacteria. As we shall see later, however, a fallacy is likely to lie in this sort of experiment unless the most careful controls are prepared and observed. We know that bacteria may grow well, even luxuriantly, in weak dilutions of many salts, but at the same time become fairly resistant to some but not all lytic agents. The cultural transformation involved in these cases is the change from the S to the R form of culture, or to some other resistant cyclostage.

In many of these instances it seems to me that a fallacy may easily be involved in the interpretations; and this concerns the action of the substance on the cultural substratum as opposed to its action on the bacteriophage itself. D'Herelle has already called attention to this point, although without grasping its full significance. If lysis does not occur in a tube containing bacteriophage, sensitive culture and foreign substance, the results may be due to either of two things. First, the bacteriophage may itself be affected adversely by the substance; second, the substance may act on the bacteria causing a transformation to a type possessing greater resistance to the bacteriophage being used, usually in the direction of the R form. We already know from our studies on enforced dissociation that growing bacteria in the presence of certain salts serves to transform them in the manner indicated previously; and that the R type culture is more resistant than the S type to the original lytic stimulus, though not necessarily to all lytic stimuli. D'Herelle [193] has already shown the effect of nitrate and acetate of lead, and of nitrate and sulphate of silver, in favoring the development of secondary, resistant cultures arising under the lytic stimulus. In this connection the important point that has escaped him is that certain salts (potassium permanganate, cadmium nitrate, potassium bichromate, mercuric and lithium chloride, or even high concentrations of sodium chloride),* and presumably also the salts which he has mentioned, added to normal, sensitive culture, tend to enforce a dissociation in the direction of the resistant form (R type) which is similar to, if not identical with, those secondary cultures produced under the influence of the "large" or a type bacteriophage. Either salts alone or the bacteriophage alone manifestly serve to turn the trend of culture development into the same path, and this is represented by the S to R transformation. It may be added that this is not true of salts alone, but of many other substances, both organic and inorganic (see Microbic Dissociation,[168] section 11).

With reference to the effect of bile, Hauduroy recently made an interesting contribution. This substance was found to arrest the lytic reaction although it did not destroy the principle. Hauduroy believed that in the gallbladder the action of the bile may be to inhibit the

* See microbic dissociation,[168] section on "Incitants to Dissociation" for detailed treatment of this subject.

activity of the bacteriophage, and so to conserve the typhoid bacteria. He also suggested the same possible action in the case of bile mediums commonly used for culturing B. typhosus. In these and in similar instances it seems more probable that the action of the bile is not on the bacteriophage, but on the culture itself, causing it to undergo at least a partial transformation toward the rough (R) type. This interpretation is in agreement with d'Herelle,[193 a] although he does not recognize the nature of the influence of the bile or other substances which inhibit the lytic action. He says: "Bacteriophagy is operative only in living and *normal* bacteria." It may also be borne in mind, in this connection, that typhoid bacteria, isolated from the gallbladder or from the urinary bladder in the case of carriers, are often of the R type, and often unaccompanied by the bacteriophage. Furthermore, one may remind oneself of the strong dissociating influence which bile possesses in forcing the dissociation of the sensitive S type pneumococcus into the R state, as shown by Reimann and Amoss, respectively.

Chloroform and Ether.—D'Herelle found that when the bacteriophage was kept in an atmosphere of chloroform, attenuation could be noted as early as $3\frac{1}{2}$ hours. Fejgin observed that ether extraction was borne without the loss of activity, and from this test concluded that the agent could not be a living thing. Similar conclusions were reached by Stassano and Beaufort from their detailed experiments; also by Nakashima.

Oxygen.—The effect of oxygen was studied by Schwartzmann. The availability was ascertained by the ratio of volume of medium to the surface area of the culture. He found that if the figure obtained by dividing the surface area (in sq. cm.) by the volume of medium (in cc.) was greater than 0.5 lysis and regeneration did not occur. If, however, the number was less than 0.5, lysis and regeneration occurred. Again, in 1926, Schwartzmann reported further on this point. While oxygen was found not to affect the lytic principle alone, the growth of the sensitive bacteria under aerobic conditions (exponent 3.5) for three days produced changes in the microbes whereby they gained resistance. Thus, while cultures grown aerobically at P_H 6.0, 7.0, 8.4 and 9.0 conserved their susceptibility, the cultures cultivated aerobically at 7.6 acquired resistance. At the same time, when the same strain was cultivated at P_H 7.6, but in the absence of air, it maintained its sensitivity. The question therefore arose whether the resistance of B. coli could be induced as well under anaerobic conditions. Cultures

of the susceptible strain were made in broth tubes having a range of
P_H values. Having then been covered with a layer of petrolatum, they
were grown for three days. The results showed that while the anaerobic
cultures in tubes having P_H values of 6.0, 7.6, 8.4 and 9.0 conserved
their susceptibility the culture grown at P_H 7.0 anaerobically was resistant. The author, therefore, concluded that cultivating a sensitive strain
in air at a P_H of 7.6 or in the absence of air at a P_H of 7.0, can determine a resistance of this strain to the bacteriophage.

Since the time of Schwartzmann's work d'Herelle has reported
results which though not exactly failing to confirm the observations of
the former (since his methods were different), leave the issue somewhat confusing. D'Herelle maintains that the lytic action occurs equally
under aerobic and anaerobic conditions; and that this refers to B. shigae,
staphylococcus, V. cholerae and B. pestis in tubes 17 mm. in diameter
containing 10 cc. of broth. Under anaerobic conditions the reaction
occurred more slowly. In order to demonstrate the possibility of lysis
under conditions of free oxygen supply, he used two methods: (1)
Erlenmeyer flasks with the broth 7 to 8 mm. deep and provided with
air circulation; (2) Petri dishes 14 cm. in diameter with broth 2 to
3 mm. deep. Under both of these conditions d'Herelle reported that
the lytic reaction was extremely rapid. Shiga cultures underwent lysis
in $2\frac{1}{2}$ hours; staphylococcus, V. cholerae and B. pestis cultures in
$7\frac{1}{2}$ hours. Indeed, when these tests were conducted several times in
succession, the virulence of the bacteriophages appeared to undergo
marked exaltation which was not lost when the tube method was again
employed. D'Herelle stated that this method of generating lytic principle possesses special value in the production of large amounts of filtrate
for therapeutic purposes (dysentery in the Sudan). For such a demand
he recommends the use of large crystallization dishes, left uncovered.
At present, these results cannot be reconciled with those reported by
Schwartzmann. Given at the start normal, well-established S type cultures of the common intestinal bacteria, the experience of numerous
investigators would lead to the conclusion that, for practical purposes at
least, slight differences in reaction or in oxygen accessibility play a small
rôle in the lytic reaction. It is known, certainly, that, on an agar surface freely exposed to air, lysis takes place quickly and extensively. To
this it may be added that Larkum [351] has been able to confirm d'Herelle's
results so far as they appertain to the staphylococcus bacteriophage.

Larkum observed that the bubbling of oxygen through the culture increased the strength of the lytic agent. Carbon dioxide had the reverse effect.*

Glandular Extracts and Commercial Enzyme Preparations.—These may be considered from two points of view: first, their influence in generating the lytic agent in cultures; second, their influence on the bacteriophage or on the lytic reaction. Since the first point has already been dealt with, only the latter will be considered here.

Wollman, in 1924, studied the action of trypsin, pepsin and papain. The active substances were reported as having an inhibitory effect on lytic action, while the influence was not apparent when inactivated samples were used. When a small amount of trypsin was added to Shiga cultures they no longer revealed plaques on agar. Wollman, therefore, concluded that the lysogenic factor was bound up with an albuminoid substance which was destroyed by tryptic digestion.

In further tests, however, Wollman and Mme. Wollman showed that a coli bacteriophage was not so inhibited by trypsin. But a curious feature of these experiments was that this resistance to trypsin was maintained by the coli bacteriophage after it had been adapted to the Shiga bacillus, while the susceptibility of the Shiga bacteriophage was maintained after this agent had been adapted to B. coli. In a still later publication Wollman stated that, even in the Shiga bacteriophage, there is never a complete destruction by the trypsin, even after several weeks at 37 C. The actual difference between the coli and the Shiga bacteriophage thus appears to be only in the degree of susceptibility.

The fact that trypsin has slight effect on the vitality of the bacteriophage was shown by Arnold and Weiss in 1926. These workers showed that it was not injured appreciably by tryptic digestion. They also demonstrated that the principle could be freed from the antigenic bacterial proteins by digesting lytic filtrates with prepared trypsin solutions.

The chief result of these studies is to indicate that the bacteriophage is not to be regarded as a lifeless, protein substance arising from bacterial metabolism or autolysis. Although their implications are not yet wholly clear, the results reported at least do not militate against the view that the bacteriophage may be protein matter in the living state. It is known that the proteolytic enzymes do not attack living cells.

* It may be noted here that, in the experiments dealing with this subject, no distinction has been made between the α and the β principles. Do both have the same reaction to anaerobic conditions?

Radiation.—D'Herelle, reviewing the experiments of Brutsaert on the effect of radium emanations, concluded that the results were essentially the same as on the bacteria themselves. In 1925, Bruynoghe and Mund also showed that the lytic principle, in contact with 7 to 8 millicuries, conserved all its properties. They also demonstrated that the principle did not regenerate at the expense of radiated typhoid bacteria. On the other hand, Lacassagne and Paulin, in 1925, reported that the bacteriophage was killed by radium emanations under conditions similar to those bringing about the destruction of toxins and filtrable viruses (Ferroux and Mutermilch *).

The experiments of several investigators previous to 1925 showed that the agent is sensitive to ultraviolet rays. While Gildemeister believed that the resistance was about the same as that of the bacteria, Gerretsen and his collaborators found that the bacteriophage of B. radicicola was somewhat more resistant than the vegetative cells. While the latter were destroyed in fifteen minutes, the former was still active after two hours of radiation. Biemond also showed the lethal effects of ultraviolet rays. McKinley, in 1926, observed that an exposure for forty minutes at a distance of 1 foot (30.5 cm.) was sufficient to destroy both a coli bacteriophage and a sample of Levaditi's encephalitis virus. The presence of serum from a normal rabbit protected both against the harmful effect of such radiation.

From the data available at present, it may, perhaps, be concluded that the behavior of the lytic agent under ultraviolet radiation is not out of harmony with a conception of this principle as a living agent. Indeed, its behavior, taken in conjunction with that of the bacterial cells, seems to suggest this possibility.

Temperature.—Recent studies add little to what is already known regarding the resistance of the bacteriophage to heat. D'Herelle stated that the most common point of thermal inactivation is about 75 C. acting for 30 minutes. In the case of a Shiga bacteriophage a temperature of 72 C. produced inactivation so far as the production of plaques was concerned, but such a heated principle could be regenerated to original activity by three serial passages. The attenuation of virulence may begin at a much lower point, however, sometimes as low as 62 to 64 C. (typhoid and paratyphoid). Staphylococcus bacteriophage seems to be more sensitive, judging from the report of Twort.[322] In general the attenuation begins about 10 degrees below the point at which the

* Compt. rend. Soc. de biol. 93, pp. 608 and 1226, 1925.

activity is definitely lost. In all cases, however, d'Herelle advises that it is necessary to perform several serial passages in order to ascertain whether the principle may be regenerated. In some cases, especially if many passages are involved, and the culture is sensitive, there may be a question whether the old principle is regenerated or a new one is produced. Evidence which justifies the apprehension of such a possibility has been presented in a foregoing section. To this it may be added that Hauduroy has reported the destruction of the bacteriophage in the dry state only at a temperature of 135 C. (275 F.), but this has not been confirmed.

It is a point of special interest that Koser, who obtained a bacteriophage for a thermophilic culture, producing active lysis and typical plaques at a temperature of 58 C., observed that the point of thermal inactivation was not appreciably higher than that of the Shiga bacteriophage working on its homologous substratum.

One of the most interesting observations on the heat resistance of the bacteriophage was presented by Gratia in 1923, in connection with his large area and small area strains of the coli bacteriophage. He has briefly reported that the inactivation point of the large area principle (weak) was from 60 to 62 C., while the small area principle resisted a temperature of 72 C. To this a point of possible significance not recognized by Gratia may be added at this time. Gratia showed that the large area principle was regenerated at the expense of the "diffuse" type of sensitive culture (S type), while the small area principle was regenerated at the expense of the "agglutinative" form of culture (R type). From recent work on microbic dissociation in members of the colon-typhoid-dysentery group, it is known that Gratia's "diffuse" sensitive culture is the equivalent of the smooth S type, while his "agglutinative" culture is the equivalent of the rough R type. It is known, moreover, that the dominant antigen of the S type is heat-labile, while the dominant antigen of the R type is heat-stable. From these considerations it might appear that each type of bacteriophage (the large area type and the small area type) generated, respectively, from the S type culture and the R type culture, partook of the characteristics of the respective host cells, so far as its relation to heat inactivation is concerned. Gratia was not aware of these facts at the time of his publication, but it seems to me that they are of considerable interest and significance, the nature of which will be considered later.

My student, Eugenia Dabney, has been able to confirm Gratia's observation (B. coli) on the difference in points of thermal inactivation for the principles related to the large areas and small areas, respectively, in the case of B. paratyphosus B. The large area principle (sealed in capillary tubes) was inactivated at 63 C. when heated at this temperature for 30 minutes, but was not inactivated at 60 C. The small area principle, under similar conditions of test, was inactivated at 75 C. but not at 70 C. Attempts have not yet been made to obtain narrower temperature limits. In this circumstance we may probably find the explanation of the diverse reports on the heat-lability of the bacteriophage. Twort stated that his staphylococcus principle was destroyed by heating at 60 C. for 1 hour. D'Herelle observed a partial inactivation of a Shiga bacteriophage heated at 60 to 65 C., while Kuttner found the temperature limit between 70 and 75 C. It seems possible to reconcile these various observations through the fact that the actual point of thermal inactivation will depend on whether one is dealing with a "pure," large area strain, a "pure," small area strain, or a mixture of the two.

I shall now examine the effect of temperature, not on the bacteriophage itself, but on the process of lysis, in which the physiologic behavior of the lytic agent needs to be correlated with that of the cells of the substratum.

D'Herelle [193] has already pointed out that the lytic reaction will occur, not only at the temperature of optimum growth of the bacterial species constituting the substratum, but at temperatures considerably removed from this point. Indeed, the range of temperatures within which lysis may occur is rather wide. Thus, Kuttner [225] has shown that the reaction with B. typhosus may be stronger at 41 C. than at 37 C. (98.6 F.), but that it does not occur at 45 C. D'Herelle has indicated that lysis of the dysentery bacteria may take place at any point between 8 C. (46.4 F.), and 41 C., but that the time required is variable. In B. pestis the maximum was 41 C. In staphylococcus, lysis was observed in one strain at 43 C. In B. coli, according to d'Herelle,[193] Doerr and Gruniger have reported that lysis does not occur at 43 C.: but d'Herelle has stated that, in this species, it may take place as high as 46 C., and he is inclined to place the limits between 8 and 46 C. He believed, however, that further study might extend these limits in both directions.

That the limits were actually susceptible to such extension has been amply demonstrated recently. Koser, in 1926, working with a true thermophile the optimum growth of which occurred between 45 and 52 C., but which manifested good growth between 20 and 60 C., obtained, by the use of sewage contaminated river water, according to my usual method, an active and typical bacteriophage. This principle, applied to cultures grown at 55 to 58 C. yielded plaques having a diameter of 2 to 3 mm. At 37 C. the plaques were slightly smaller than at 55 C. When the cultures were grown at room temperature the areas appeared only after 2 to 3 days. In tubes of broth, grown at 58 C. overnight, lysis was distinct. Growth at 60 C. left the results in doubt, and at 62 C. they were clearly negative. The lytic filtrate resisted a temperature of 70 C. applied for 30 minutes, but was destroyed at 75 C. in the same length of time. This principle was not active on Shiga nor on several other thermophilic species. It is a point of much interest in Koser's work that, although the optimum and maximum temperatures for growth and for lysis are so much higher than for any organisms previously reported, the point of thermal inactivation corresponds perfectly with that of lytic agents active for members of the colon-typhoid-dysentery group. Koser presents the interesting view that, if one looks on the bacteriophage as a foreign filtrable virus parasitizing the bacterial cells, the results obtained with the thermophilic strain must lead one to postulate the existence of a thermophilic filtrable virus. The author, however, does not commit himself to any definite view from the results of his important experiments.

Just as Koser was able to extend the upper temperature limits of bacteriophage action to 58 C. Elder and Tanner have recently extended the lower limits of the reaction. Working with an unnamed psychrophilic species which grew well at 4 C., but not at 37 C., they have shown that the bacteriophage obtained operates at 4 C. or below this point.

In analyzing the significance of these important investigations, it seems to me that they carry definite implications, namely, that any pure-line bacteriophage partakes of the specific growth characteristics of the bacterial species on which it acts, and from which it arises. Other evidence supporting this point has already been mentioned in connection with Gratia's large area and small area principles. With these facts in mind, the bacteriophage might easily be regarded as a stage in the life history of the species represented by the sensitive substratum. These observations are definitely opposed to d'Herelle's conception of the strict unity of the bacteriophage, unless it possesses a much more diversified and plastic structure than is usually believed to be characteristic of living things.

In concluding this discussion of the influence of heat on the bacteriophage, and of temperature on the lytic process, I may say that, all data considered, the fact is strongly suggested that a labile or perhaps a living thing is being dealt with, the heat resistance of which lies in an intermediate position between that of the sensitive vegetative cells and that of the resistant spores.

Viscosity of Medium.—A number of investigators beginning with Doerr have shown that the addition to a culture medium of such substances as gelatin or tragacanth served to inhibit the lytic reaction. Nakamura showed that some bacteriophages, but not others, were inhibited by gelatin and similar substances. Brutsaert made similar observations. Hauduroy was the first to present the commonly accepted explanation of this action, namely, that such substances increase the viscosity but do not operate through any chemical property. After confirming the tests of Hauduroy, d'Herelle concluded that the inhibition is due merely to a reaction between colloids.

The failure of plaques to appear on gelatin mediums as reported by several investigators has been ascribed by d'Herelle to the circumstance that the gelatin is but slightly permeable to the products arising in the course of growth of the bacteria. He found that when the layer of gelatin was very thin and superimposed on a layer of agar typical plaques appeared. While this may be true, I believe that we should hold ourselves open to the view that the cultivation of normal, sensitive cultures on gelatin may accomplish some cultural modification which is associated with the acquisition of greater resistance to the lytic agent. At present little is known regarding the influence of gelatin mediums on the dissociative reaction in general, or the S to R transformation in particular. It is known, however, that certain culture types exist which will develop in broth but not on agar, and of others that will develop on agar but not in broth. There may easily be other stages that will develop on agar but not on gelatin. I cannot yet accept the fact that the current explanations of the inhibiting rôle of gelatin on lysis, or on plaque-formation, represent the whole story. Indeed, if I recall correctly, some investigator has reported that, although the organisms growing in gelatin do not seem to undergo lysis, there is still an increase in the number of lytic corpuscles.

Conclusions.—In summarizing these recent reports on the influence of various substances or conditions of growth on the bacteriophage, or on the lytic reaction, it must be admitted that evidence cannot be found in them which bears conclusively on the nature of the agent concerned. In the majority of the reports cited it seems more likely that the interpretation of the reaction has been determined by the influence of the substance on the sensitive bacteria than that it has concerned an influence on the bacteriophage itself. D'Herelle has well said that most of such instances can be summarized simply in the statement that "bacteriophagy is operative only with living and normal bacteria."

But what are normal bacteria? By "normal" d'Herelle means the typical culture which grows well on artificial culture mediums and is sensitive to the bacteriophage. A form of culture that is resistant to a bacteriophage which is able to attack another form of the same culture is, for him, not normal. Thus, d'Herelle implies that a cyclostage that is not susceptible to lytic action cannot be normal. Such a view cannot be accepted. In an earlier publication [168] I have presented the grounds for questioning the common meaning of the term, normal, when applied to bacterial cultures. New knowledge of the dissociative reactions compels the view that any bacterial culture cannot be regarded as homogeneous, but as made up of various cyclostages (Enderlein,[117] 1925), any one of which, under favorable conditions of growth, may come into such a degree of prominence in the cell population as to represent the dominating type. Among these various cyclostages, each is just as "normal" as any other. The R cell type is not abnormal merely because it has lost, perhaps, motility, capsules, virulence or susceptibility to one lytic agent. The O type culture is not abnormal because it presents a different colony form, different fermentation reactions or a different antigenic configuration from that of the smooth S type culture. Moreover, the smooth S type culture is not "normal" merely because it is often the form that grows most readily on artificial culture mediums, and is conveniently susceptible to bacteriophage action. All of these forms of culture are strictly normal; and, at some time or other, arise in any bacterial culture, unaided by any artificially prepared bacteriophage filtrate. If we are to accept the view that the bacteriophage is a foreign filtrable virus, then we must be prepared to say that it possesses a dual nature and is a virus that can multiply at the expense of only certain of the cyclostages in the developmental history of the species. This might be possible, but it is hardly probable. At least, there exists another explanation soon to be presented.

6. THE INFLUENCE OF THE BACTERIOPHAGE ON ITS SUBSTRATUM

From the voluminous literature that has gathered about the bacteriophage in the past seven or eight years one is in danger of concluding that its chief influence is to effect the actual destruction of bacterial cells by a process of lysis, and that the most important correlated effects are the inhibition of bacterial growth and the production of areas of lysis in cultures grown on solid culture mediums. If, on the other hand, secondary, resistant cultures are produced, or if the bacteriophage seems to come into equilibrium (Bordet) with the cells, perhaps maintaining itself in the culture by virtue of a sort of symbiosis (d'Herelle), most investigators would doubtless regard these phenomena as matters of secondary importance. The ultravirus has to live and propagate. If it can do so without the actual destruction of the cells, or with the destruction of only a part of them, that is immaterial to the bacteriophage. But, if it is highly "virulent," it does not permit escape. That is one point of view, from which the agent is regarded as an invading and hostile element.

But, fortunately, it is not the only point of view which permits us still to retain the hypothesis that the lytic agent is a living thing. One must bear in mind that populations of organic beings, especially when maintained under artificial conditions, are themselves often self-destructive. Societies, usually well regulated, may sometimes experience almost complete annihilation merely because certain apparently lawless elements, comprising normally but a small fraction of the population, are, as a result of some disturbing environmental factor, developed into a degree of ascendancy. Such a state of the society might be termed "pathologic"; but it is pathologic only because of certain intrinsic maladjustments among its members—not necessarily because it is invaded from without by disturbing elements. There is ample reason for assuming, as Kuhn has done, that an analogous condition may arise in bacterial populations under the stress of artificial treatment, and especially under the conserving influence of repeated filtrations of bacterial cultures. Indeed, there are many reasons for suspecting that this may be the case—provided, of course, one has left behind ancient monomorphic conceptions regarding the nature of bacteria and the modes of bacterial reproduction.

This conception of the matter will be developed more fully in a later section, but the subject is introduced at this point for the purpose of presenting the point of view that it is not necessary—indeed that it may be actually disadvantageous—to focus one's attention too closely on the lytic features of the bacteriophagic reaction. As I have pointed out in an earlier section, the most significant aspect of bacteriophagic action is the rapid enforcing of microbic dissociation on the culture concerned; the transformation of some culture elements and the apparent elimination of certain others. The essential point which has until recently escaped the attention of most bacteriologists is that culture modification is the primary characteristic of transmissible autolysis; and that the destruction (or, let us say, disappearance) of certain cells is a matter of secondary importance, a point that has been emphasized also by Hoder and by Breinl. In all of d'Herelle's writings, and in those of most others, chiefly the phenomena of lysis, inhibition and plaque-formation have been emphasized, although the production of "muta-tions" is given a place. The trend of recent studies, however, makes it increasingly clear that an understanding of the lytic process or of the mechanism of the reaction will not be reached until these aspects of the phenomenon are tentatively put aside, and the attention turned to the intimate biology of the sensitive culture, its reproductive mechanism and the remarkable series of transformations which are constantly passing before the eyes of one whose mind is not too fully prepared for anticipated eventualities.

REFERENCES

Variation and "Mutation" Accompanying the Lytic Reaction: d'Herelle;[192, 193] Bordet;[40] Fejgin;[134, 126] Flu;[131] Gratia;[148, 149, 159] Eastwood;[112] Petrovanu;[276] Busson and Ogata;[79] Breinl and Fischer;[50] Breinl and Hoder;[51] Hoder;[197] Kasarnowsky;[207] Manniger;[241] Arkwright;[9] Hadley;[168] Nobechi;[262] Béguet;[31] Wollman;[335, 336] Grumbach and Dimtza;[359] Sonnenschein.[270]

Production of Secondary Cultures: d'Herelle;[188] Bergstrand;[32] Bruynoghe and Maisin;[73] Gratia;[148, 149, 150] Schnabel;[300] Breinl and Hoder;[51] Hoder;[197] d'Herelle and Hauduroy;[196] Hauduroy;[177] Kimura;[212] Bronfenbrenner and Korb;[57] Bronfenbrenner, Muckenfuss and Korb;[59] Wollman;[335, 336] Fejgin;[134, 126, 128] Fejgin and Lararevitz;[128] Kasarnowsky;[207] Grumbach and Dimtza.[359]

Characteristics of Secondary Cultures:

Morphology.—d'Herelle;[168] Bergstrand;[32] Bruynoghe and Maisin;[73] Gratia.[148]

Capsules.—Caublot;[86] Hadley;[164] Kimura;[212] Sonnenschein.[315]

Fermentative Reactions. — Hauduroy;[174, 177] Ogata;[263] Bergstrand;[32] Fejgin;[131] Grumbach and Dimtza.[359]

Toxicity.—Fejgin.[124]

Relation to Hemolytic Influence of Staphylococci.—Epstein and Fejgin.[218]

Virulence.—d'Herelle;[193] Davison;[104] Bordet;[38] Gratia;[148] Fejgin;[124] Bronfenbrenner and Korb;[57] Bronfenbrenner, Muckenfuss and Korb;[59] Hadley;[168] Béguet.[31]

Manner of Growth.—d'Herelle;[193] Hauduroy;[176, 177] d'Herelle and Hauduroy;[196] Gratia;[148, 149] Wollman.[335]

Symbiosis.—See foregoing.

Types of Resistant Cultures.—Gratia;[155, 156] Breinl and Hoder;[51] Bronfenbrenner, Muckenfuss and Korb;[60] Hadley and Dabney.[169 a]

Relation to the Two Types of Lytic Agent.—Gratia and de Kruif;[161] Gratia;[155] Hadley and Dabney.[169 a]

Filtrability.—d'Herelle;[188] Hauduroy;[176, 177] d'Herelle and Hauduroy;[196] Fejgin.[223]

Reversion to Sensitivity. — d'Herelle;[183] Bruynoghe;[98] Arkwright;[9] Nobechi.[282]

Serologic Aspects.—See section on serologic reactions.

Relation to R Type from Normal Dissociation.—See section 14 in Microbic Dissociation.[168]

Use of Differences for Diagnostic Purposes.—Grumbach and Dimtza.[359]

DISCUSSION

Secondary Cultures; Mode of Production.—D'Herelle was the first to point out that, although the addition of a strong bacteriophage to a sensitive culture may effect its apparent sterilization, it often happens that all of the bacteria are not destroyed. Some, successful in "acquiring a resistance" to the lytic agent, seem to become the progenitors of a resistant group which is represented by the secondary growth, differing in many respects from the primary culture. The same sort of secondary growth may appear on the surface of a solid medium, after the primary culture has disappeared.

But d'Herelle also noted that all the resistant strains arising in a single tube might not be alike. Some were resistant without carrying the lytic agent, while others might harbor the principle for many culture generations. The latter were early recognized by Bordet and termed lysogenic forms. In these d'Herelle assumes that the lytic agent is living in a state of symbiosis. These variations have been described by many investigators, and for many species. There has been a tendency to observe a greater variety among these new types than was first indicated. This is shown particularly in the work of Gratia who recognized 11 different culture forms arising in a coli culture under the influence of the lytic principle. Several were also pictured by Bergstrand in 1922, by Fejgin for B. proteus in 1924, by Hauduroy for Shiga in 1924 and also by Breinl and Fischer, Breinl and Hoder, and Hoder alone at a later date. As Bruynoghe and Maisin indicated in 1921, similar secondary, resistant forms arise in cultures of staphylococcus. That some of these resistant types may develop in Chamberland filtrates of lysed cultures has been shown by d'Herelle, Hauduroy, and d'Herelle and Hauduroy; also by Fejgin for B. proteus and B. typhosus. The results of d'Herelle and Hauduroy present to us the somewhat remarkable picture of the filtrable virus (bacteriophage) living in a state of symbiosis with a resistant, filtrable and ultramicroscopic form of the organism of the substratum.

In a few cases, secondary cultures differing in certain respects from the usual resistants have been observed. Thus, d'Herelle [193b] mentioned a mucoid resistant in the case of Shiga, and had some difficulty in explaining its appearance. Kimura and Sonnenschein also referred to mucoid cultures arising from lysis. Not all of Fejgin's resistant Shiga or proteus forms were alike. Gratia pictured several types of B. coli departing radically from the usual resistant type. Breinl and Fischer, also Breinl and Hoder, observed that the lysis of B. paratyphosus B yielded four different variants. They clearly noted the relation of these forms to those commonly seen in the course of "mutation" reactions in bacteria.*

Significance of the Secondary Cultures.—As to the actual origin of these forms, which may be few or numerous, depending on the "strength" or type of the lytic agent, the nature of the culture substratum and other factors, d'Herelle concluded that they are "the result of an adaptation undergone by the bacterium which acquires an immunity to its parasite." "They have their origin in the phenomenon of natural selection whereby some bacilli show a greater aptitude than others to the acquisition of resistance to the bacteriophage." For d'Herelle this view is based on the assumption that, in general, all the cells of a given culture are alike—that is, aside from the ability to acquire resistance. As he says at one point, "A pure bacterial strain is not subject to transformations." This statement stamps his work with the seal of monomorphism.

Gratia,[149] in 1921, was one of the first to question d'Herelle's conclusions on this point. This author held that resistance was already present in certain cells of the culture, prior to the action of the bacteriophage; in other words, that the pure culture was composed of differentiated cells, some sensitive, some more or less resistant. In his latest book d'Herelle [193] combatted this view. Gratia's conclusions were, however, supported by the observations of Breinl and Hoder and of Hoder in 1925, and found their actual explanation in the much earlier observations of Arkwright in 1921. F. M. Burnet,[77] moreover, in 1925, concluded that resistance depends on some variable factor in the bacteria, rather than on variations in the lytic principle. He found, as Gratia, Zdansky and one or two others had pointed out, that bacteria resistant to one strain of bacteriophage might be susceptible to another strain. Resistance was sometimes correlated with the appearance in the culture of filamentous forms of growth (coli). Bronfenbrenner and Korb and Bronfenbrenner, Muckenfuss and Korb in 1926 reached conclusions agreeing with those of Gratia.

---

* I wish here to make acknowledgment of the omission of the significant study of these authors, in my earlier work on microbic dissociation; only recently did I come on their paper.

Wollman in 1925 could produce resistant cultures with attenuated, but not completely inactivated, lytic filtrates, and concluded that one cannot choose between the hypothesis of selection and the appearance of new characters in the organisms, but that the nature of the modifications favored the latter view.

The actual mechanism involved in the production of the resistant cultures is apparently somewhat intricate, and from this point forward, it will be observed that the phenomena of the bacteriophage enter more and more into the field of microbic dissociation. For this reason, I shall pause to consider several fundamental points dealing with the dissociative reaction, a general review of which I have presented in an earlier publication,[168] and will refer to in greater detail in section 10.

Although the differentiation of the cells of a pure-line culture into components differing markedly from one another was observable in a considerable number of earlier reports, Arkwright,[8] in 1921, was the first to present clearly the significance of the subject, and to demonstrate the possibility of dividing cultures of several common bacterial species into at least two distinct components, the smooth (S) and the rough (R) types. These two forms, arising in pure-line cultures, differed in colony form, growth in broth, biochemical and serologic reactions and virulence. The transformation from S to R occurred easily, while the reversion of R to S was effected with greater difficulty. These observations are manifestly in opposition to d'Herelle's view that a pure-line strain is not subject to transformations.

In 1921, Gratia[150] pointed out that, of these two culture types (S and R), the former was highly sensitive to the bacteriophage while the latter was more resistant. Later, in the same year, he also demonstrated the wide variety of culture forms that could be produced by permitting bacteriophage of different strengths to act on a sensitive culture. Although not realized by the author at the time, these results concerned in a striking manner the dissociation phenomena manifesting themselves in the coli culture studied, and demonstrated at least the partial similarity between the R type arising from simple dissociation and that arising under the influence of the weak lytic agent. He thus demonstrated that resistant cultures do not owe their origin, exclusively at least, to the action of the bacteriophage, as d'Herelle has maintained.

The correspondence between normal variations and those variations appearing in the course of action of the bacteriophage was also noted by Eastwood in 1923, and further by Breinl and Hoder and Hoder in 1925. These workers observed the similarity in the variants of members of the colon-typhoid-dysentery group, regardless of whether they occurred in old cultures, in cultures submitted to the influence of unusual temperatures or in cultures exposed to the lytic principle. The differences observed concerned colony form, growth in broth, sometimes serologic reactions and resistance to the bacteriophage. Fejgin, in 1923, also noted in cultures of Shiga the spontaneous appearance of three variants which were resistant to the bacteriophage. Petrovanu, while studying experimental cholera infections, observed that the inoculated vibrios went through a rapid series of transformations in the body of the animal, and that the cultures taken yielded two new races. One of these was susceptible to lysis, while the other was resistant. Moreover, Busson and Ogata, in 1924, reported data indicating the apparent identity of the resistant cultures (R) arising from normal dissociation and the resistant strains (SR) arising from lytic action. In 1926, Sophie Kasarnowsky reported an important study in which she showed that a culture sensitive to the lytic agent can be made resistant merely by growing it in the filtrates of cultures that had already become resistant. Further, that bacteria made resistant in this manner could be differentiated both culturally and serologically from the original strain. They were, moreover, wholly analogous to the resistant cultures produced under the lytic influence. Some of the details of the studies of this author remind one of the interesting results obtained by E. Burnet [76] by growing cultures of M. melitensis in filtrates of paramelitensis cultures. In 1926 Manniger also reported that in 41 coli strains cultural modifications were always observed and resembled those attributable to the action of the bacteriophage.

These and similar examples that could be cited are sufficient to demonstrate that d'Herelle is plainly in error when he states that secondary, resistant cultures are always the result of bacteriophage action, at least, as he understands and depicts the reaction. To support his arguments, he contends that, wherever such cultures are found, indications show that they have been produced at some time in the past by the culture in question having come in contact with the lytic agent. The truth of the matter is, however, that any sensitive culture, free from lytic manifestations, can be transformed into a resistant strain by means of a variety of simple laboratory procedures such as those I have referred to in detail in a previous paper.[168] Although d'Herelle has recognized some sort of a "modification" of bacteria under the influence of salts, serum, etc., whereby they become "abnormal" and resistant to lysis,[193c] he has failed to recognize the exact nature of the change involved. This failure to appreciate the normal and spontaneous origin of resistant cultures, or the possible origin of such cultures under the influence of artificial incitants, is apparently due to the circumstance that his entire outlook on the biology of the bacteriophagic reaction has been obscured by his firm adherence to the dictates of the old monomorphic conception of the nature of bacteria. He has failed to see that bacteriophage action is an intensified microbic dissociation.

What, then, is the truth of the matter with reference to the controversy between d'Herelle and Gratia regarding the origin and significance of the resistant, secondary cultures? In answer to this question I believe it may be said that in a measure both are right. There cannot be any doubt, as I have proved in many still unpublished experiments, that d'Herelle is correct when he states that the addition of lytic filtrate to young sensitive cultures may produce resistant forms. He is mistaken only in believing that such resistant forms are produced exclusively under the bacteriophagic influence. Gratia is also correct when he states that resistant organisms may exist preformed among the cells of the sensitive substratum. He is in error only when he intimates that the bacteriophage does not have power to produce these resistant forms de novo in a sensitive substratum. In reconciling these two views it therefore, may be briefly stated that the bacteriophage is able to accomplish quickly and in large measure what the normal sensitive culture, under usual conditions of growth, is able to accomplish for itself, in a small way and over a much greater period. All sensitive cultures, in aging, or when submitted to the influence of certain substances or environments, can easily be shown to transform themselves from the normal, sensitive culture into the rough form, or into another resistant type (R^2), in which they seem to be identical with the series of resistant cultures arising under the lytic influence. This transformation in type from S to R is one of the fundamental aspects of the normal dissociative reaction; and it is not less so, merely because it can be hastened by the stimulus imparted by the lytic principle in the form of artificially prepared filtrates.

These considerations will perhaps make my meaning clearer when I stated earlier [168] that the bacteriophage action is merely one aspect of the larger problem of microbic dissociation—a point which will be developed further, with certain supporting data, in a subsequent section of this work. Both of these phenomena involve the disappearance of one form of culture and the slow or rapid generation of new culture types which, for a time at least, are not easily susceptible to further transformations.

In them the strain has become, to a degree, stabilized in a new cyclostage, endowed with characteristics and potentialities quite different from those of the mother culture. It is also, for these reasons, that I have come to regard the transformations in culture type, occurring in the bacteriophagic reaction, as possessing a significance much greater than that of the simple destruction of certain cells by lysis, or of the generation of mutations (in the exact meaning of the term) by a foreign virus. More conclusive evidence supporting these conclusions must await the presentation of other data included in a later section of this work (section 10).

Characteristics of the Secondary Resistant Cultures.—For reasons pointed out previously, considerable interest attaches to the characteristics of these resistant culture types arising as secondary cultures after lysis, or at least appearing in the course of the lytic reaction. With respect to the morphology of the cells, as d'Herelle first pointed out, and as has since been confirmed by many other workers for a considerable number of species, the cells of the extreme resistant type are usually shorter and plumper than the cells of the mother culture. If they arise in a culture that has undergone lysis, but which has not been filtered, they are microscopic, but when they arise in filtrates (Berkefeld or Chamberland) one may justifiably assume that, in a certain point of development at least, some of them possess an ultra-microscopic stage. These points relating to resistant filtrable forms have been ascertained mainly from the studies of Hauduroy and of d'Herelle and Hauduroy, and relate to members of the intestinal group, staphylococcus and the cholera vibrio. In some cases the resistant culture has been observed to develop first in the form of minute granules which produce merely an opalescence in the culture medium. In this granular state all the cultures studied seemed to meet on a common morphologic footing, whatever their original nature,—rod, coccus or vibrio. From this granular stage, moreover, the cells of characteristic morphology often developed in time. These cells were resistant to lysis and yielded an agglutinative form of growth in broth. According to Gratia, however, some of the resistant cultures give homogeneous and diffuse turbidity rather than an agglutinative growth. These probably concern what I have earlier termed an "absolute" or "extreme" resistant type. As we shall see, however, their resistance is, in a measure, relative.

Accompanying the changes in the morphology of the cells occurring in secondary cultures are also to be observed changes with respect to capsules and flagellar apparatus. These changes are characterized by a progressive loss. In the secondary cultures of Friedländer's bacillus Caublot demonstrated the absence of capsules, and I have been able to confirm these results; also to establish the same point for Bact. ozenae and Bact. rhinoscleromatis. Here it may also be noted that, as I [164] reported earlier, the dissociative transformation of Friedländer's bacillus from the S to the R form is characterized by the loss of capsules; and the same is true of the pneumococcus, M. tetragenus, B. avisepticus and a number of other typically capsulated species; also for B. coli, as shown by Smith and Bryant. Some writers have stated that the presence of capsules is an obstacle to lysis by the bacteriophage, but the results obtained by Caublot, also by myself, on a heavily capsulated form serve to dispel this illusion. In 1925, Kimura first showed that cultures undergoing lysis might produce slimy secretions which increased the viscosity of the medium, although these changes were not due to capsules. Bronfenbrenner later made a similar observation. Recently, Sonnenschein [315] has considered the subject of slimy paratyphoid cultures in relation to bacteriophage action.

To the foregoing may be added that the transition from the sensitive to the resistant type of culture is commonly accompanied by the loss of flagellar apparatus. This occurs in the colon-typhoid-dysentery group, but has also been noted in other species, such, for example, as B. proteus (Fejgin [126]). Here it may be recalled that a similar loss of motility usually accompanies the transformation from S to R type culture in normal dissociation, although Arkwright has mentioned certain exceptions.

Biochemical Reactions.—That the secondary, resistant cultures manifest different fermentation reactions from those of the sensitive culture has been shown by many investigators. In B. pyocyaneus I [163] have shown that the R form has lost all power to produce pyocyanin, although it still generates fluorescin. Its proteolytic activities are also markedly diminished. On the other hand, some cultures, as B. proteus (Fejgin [126]), take on a yellow chromogenesis, as is also characteristic of the R dissociate of many species. Changes in fermentation reactions have been noted many times (Bergstrand, Hauduroy, Fejgin, Ogata, Grumbach and Dimtzer and others).

Virulence and Toxicity.—The impression gained from the latest book of d'Herelle is that the resistant cultures possess greater virulence and are to a smaller degree phagocytable than is the case with the normal, sensitive cultures. This conclusion was supported by evidence on B. pestis presented by d'Herelle, and by reports of Bordet and

Ciuca and by Gratia on B. coli and by Davison on the Shiga bacillus
In addition, d'Herelle was of the opinion that the failures to secure
favorable results from the administration of the bacteriophage in cases
of typhoid were due, in some instances, to the circumstance that the
organisms became resistant to the lytic agent, and, therefore, more
virulent. That this situation is not general, however, and that it may
even be the reverse of the general rule is indicated in a number of
instances in which the quality of virulence in the secondary cultures
has been carefully studied. For example, Fejgin showed that her
secondary Shiga cultures lacked not only virulence, but also toxicity.
Bronfenbrenner and Korb and, later, Bronfenbrenner, Muckenfuss and
Korb showed that the resistant form of B. pestis caviae was lacking in
virulence. Here they observed two sorts of resistant cultures, one
giving a homogeneous or diffuse growth in broth, the other an agglutinative growth. The latter (R^1?) was nonvirulent and reverted to the
virulent type after a few daily passages in broth, while the former
(R^2?) remained nonvirulent even after 120 daily passages. These
results, which much resemble those presented earlier by Gratia, so far
as the cultural features of the two forms of resistant culture are concerned, were also in accord with the general observation that the
analogous rough cultures from normal dissociation are commonly nonvirulent, or at least much less virulent than the smooth sensitive types.
This is true for B. typhosus, B. dysenteriae, B. paratyphosus B, B.
cholerae suis, B. enteritidis, B. coli, Bact. lepisepticum, Bact. pneu-
moniae, B. pyocyaneus, the pneumococcus, certain streptococci, B. diphtheriae and several other species. All data considered, the evidence
thus seems to be opposed to d'Herelle's view that the secondary cultures
from lysis are more virulent than the original sensitive forms. That
the resistant cultures commonly possess greater vitality in the face of
certain unfavorable environments can scarcely be doubted; but that they
are thereby qualified as more effective instigators of disease is a conception one can scarcely entertain. In reality, this is the situation most
to be desired. If the power of the bacteriophage is not sufficient to
eliminate, in their natural field of action, the virulent, smooth S type of
culture, it is certainly of advantage that the microbes that remain
should be transformed into a nonvirulent and easily phagocytosed
form, rather than into a more virulent and nonphagocytable form. As
I have pointed out earlier,[168] the fact that the end-products of the dissociative reaction, whether brought about spontaneously, forced by
appropriate incitants, or determined by the bacteriophage, are commonly

nonvirulent or less virulent than the original culture form, is a circumstance possessing considerable importance in the conception of the mechanism of antibacterial immunity, as will later be observed.

Classification of the Secondary Resistant Cultures.—In considering these resistant cultures thus far, the attempt has not been made to differentiate them among themselves, although it has been indicated, from time to time, that certain differences existed. The knowledge of this matter is still deficient, but a certain sort of division may be made From the studies of d'Herelle and of Bordet, in particular, it has become apparent that there are at least two forms of resistant culture. One is resistant without "carrying" the lytic agent, and thus lacks the ability to incite the lytic reaction in sensitive cultures. This is the so-called nonlysogenic resistant. In time it may return to the sensitive state. D'Herelle, however, has been of the opinion that some cultures of this sort represent permanent mutants, but this is much to be doubted. It must be borne in mind that the R dissociates (from normal dissociation) are also often highly tenacious of their new characters, and for this reason have often been regarded as mutants, but that the reversion has usually been observed, although sometimes only after months or years of observation.

The second type of resistant culture is one which, although apparently not influenced by the lytic agent which it contains, is still able to effect the lysis of sensitive cultures. This may be accomplished either by adding some of the resistant culture as a whole, or by adding some of its filtrate. Either will precipitate the lytic reaction which, from then forward, is transmissible in series. Thus, there are at least three culture types that may be considered in their relation to lysis: the sensitive, the nonlysogenic resistant and the lysogenic resistant; of these, it is only the first that commonly yields the most characteristic plaques.

But to these one other culture type must be added; and it was the good fortune of my student, Klimek, to obtain this form from an old laboratory strain of B. typhosus. This new type, which as yet has not been fully studied, is apparently a "semi-resistant." It is lysogenic but also characterized on agar slants by the presence of minute areas of lysis. In this case, however, the areas, as well as the culture, can be perpetuated in series on agar; and in this respect the culture differs from the two resistants previously described. This form of culture furnishes a striking analogy with the cultures of B. pyocyaneus which I [163] have previously described as being both lytic and lysogenic. The fact that a form of culture bearing the same distinguishing features, but at the same time yielding a bacteriophage of the typical sort (through producing only minute areas), has now been discovered in B. typhosus, furnishes additional evidence that the lytic transformation which others and I have described for pyocyaneus represents, perhaps, the only mode of bacteriophage reaction possible in this species. Larkum [351] has made a similar observation for Staphylococcus aureus.

Aside from the characterization of the secondary, resistant cultures, therefore, as (1) lysogenic, (2) nonlysogenic and (3) lytic and lysogenic, there remains the possibility of distinguishing the resistant form which gives agglutinative growth in broth and the form that yields a diffuse growth in broth. The knowledge of these two types depends almost exclusively on the data furnished by Gratia and by Bronfenbrenner, Korb and Muckenfuss. According to Gratia, the former is characterized by a colony form which is identical with the R type from normal dissociation (large, irregular, flat, rough) and is produced by the action of the large area (alpha) bacteriophage on the sensitive S culture. It, in turn, is susceptible to the influence of the small area principle. Gratia's second type, produced by the action of the small area principle, is characterized by small, round, glistening colonies, which seem to resemble those resulting from the action of a strong bacteriophage. It is these colonies that give the diffuse growth in broth, similar to the growth of the sensitive culture. According to the evidence of Bronfenbrenner and collaborators, their resistant form of B. pestis caviae which gave the agglutinating growth in broth reverted easily to the sensitive type, while the resistant form which gave diffuse growth was a permanent variant. From these observations one may conclude that, as pointed out by Hoder, the sensitive culture acquires its resistance by degrees; and that, if the process is interrupted, resistant types are found which vary in their power to hold the new characteristics. Manifestly, the large, rough colony resistant (R^1) stands nearest to the original sensitive culture type, while the small, smooth colony resistant (R^2) is much more firmly fixed. It is not at all surprising that this should be the case with resistant cultures arising from bacteriophage action, for there is ample proof that the same train of circumstances follows the acquisition of resistance during the transformation from the sensitive S to the resistant R in the course of normal dissociation. In this reaction, also, the "extreme" resistant, or the "absolute" resistant, as it has sometimes been termed, seems to appear as the final product of the dissociative reaction. These observations indicate once more how closely the two phenomena, bacteriophage reaction and dissociation, parallel each other in the nature, origin and characteristics of the resistant types.

Conclusions.—From the data presented in this section, I believe it may be accepted that the phenomena of bacteriophage action which have been emphasized most commonly—namely, inhibition of growth, lysis of cells and formation of plaques—represent only a small part of the effects produced by the lytic agent, and probably the least significant. The important results of its action are, rather, the liberation or intensification, in the culture of the substratum, of a dissociative mechanism which serves to effect cell and culture transformations of far-reaching significance. Old forms of culture disappear, and new forms are produced. Some of these new forms both d'Herelle and Bordet have spoken of as mutants, but it will be seen later that this cannot be true. These new culture types differ in many respects from the original, and they also differ among themselves. The detailed biology of the new types, their exact mode of origin and their essential differences, however, remain to be ascertained.

7. THE SEROLOGIC ASPECTS OF THE BACTERIOPHAGE REACTION

The present status of the serologic aspect of the bacteriophage reaction is marked by considerable uncertainty and confusion. The problems concerned are particularly difficult by reason of the circumstance that the bacteriophage cannot be observed alone and isolated in its reactions. It must always be studied in relation to its substratum; and, because of this, it is of importance to be able to distinguish between influences that act on the bacteriophage itself directly, and those that act on the bacteriophage indirectly by modifying the bacterial cells. It is clear that this is also true for the influence of the chemical and physical environments surrounding the reaction; but, owing chiefly to the strong dissociating power of certain antibacterial serums on sensitive cultures, special care in the evaluation of the effects is necessary in a study of the serologic reactions.

Much of the unclear vision that has characterized work in this field has been due to the common acceptance by bacteriologists of the views of d'Herelle to the effect that the generation of resistant cultures (mutants) is due exclusively to the action of the bacteriophage or to the products of lysis. If this were true, one would not need to take into consideration the possible effect of other influences in determining the generation of resistant culture types.*   But, unfortunately for many experiments which have been conducted on this basis, it is not true; and for this reason many of the conclusions drawn are not of much value.

In all studies bearing on the antigenic nature of the bacteriophage, the phenomenon of dissociation must first be taken into account, and effectually controlled, in the experimental reactions. This point of view has usually been lacking. From the monomorphic point of view of the majority of bacteriologists who have concerned themselves with these aspects of the problem, the culture of the substratum (aside from still unexplained cultural "modifications," due to fresh isolation, aging or previous contact with bacteriophage) has been regarded as a stable indicator reagent, capable of registering faithfully, at any time, the contaminating presence of the bacteriophage. Such a conception of the test culture has exerted a purely harmful influence; for it has failed to make possible an understanding of certain facts without the recognition of which experimental endeavors in this field are of little value. These points involve an understanding that the substratum culture is not by any means a fixed, immutable thing, but a flowing, changing protoplasmic mass of highly heterogeneous nature, in which the cells are undergoing spontaneously, or under the influence of slight environmental changes, orderly transformations of great significance in relation to their susceptibility or resistance to the lytic agent. Evidence for this was provided in an earlier paper. It is sufficient to state here that, in all experiments devised to elucidate the influence of certain factors (serologic or other) on the bacteriophage, one must be scrupulous to eliminate their possible influence in modifying the cyclostage of the organism concerned, and in this way giving pictures of inhibiting actions which might erroneously be ascribed to an influence exerted on the bacteriophage itself.

REFERENCES

Antilytic Serums:

Bordet and Ciuca[43] (noted existence of).
D'Herelle (confirmed).[195]
D'Herelle and Eliava[195] (bacteriophage not destroyed by antilytic serum).
Prausnitz[288] (antilytic serum reduces number of areas, not their size).
Wollman and Goldenberg[358] (lysed culture more strongly antigenic than normal culture; antilysins not absorbed by normal bacteria; antibacterial serums have no antilytic action).
Otto, Munter and Winkler[270] (bacteriophage is antigenic).
Seiffert[300] (antilytic serum decreases size of plaques).

* D'Herelle, it is true, has stated that certain substances, such as salts, for example, "modify" the bacterium so that it is less susceptible to lysis; but he does not relate these "modified" bacteria in any way to his secondary, resistant "mutants." I have noted, however, that there is a close resemblance between the R forms produced in enforced dissociation and the resistant cultures arising from bacteriophagic influence.

Bruynoghe and Wagemans [75] (adaptation of lytic agent to antilytic serum).

Watanabe [338] (antilytic serums from "elementary" strains of bacteriophage; large and small area principles).

Gratia [155] (antilytic action of large and small area strains of lytic agent).

Arnold and Weiss [15] (action of antilytic serums resembles neutralization in toxin-antitoxin mixtures).

Kuttner [225] (failed to observe antilytic effect of antibacteriophagic serum).

Osumi [268] (antilytic serum for B. coli lytic agent; stronger inhibiting action than antibacterial serum).

Urech and Pache [333] (noted antilytic effect of antilytic serum).

Da Costa Cruz [95, 98, 100] (action of antibacterial serum is to protect bacteria rather than to neutralize the bacteriophage).

Farby and van Beneden [121] (antilytic serum neutralized by anti-antilytic serum).

Kasarnowsky and Timokin-Schuekoff [206] (action of antilytic serum is specific).

Sanderson [298] (earlier results alleging fixation by antilytic antibodies due to nonspecific fixation by rabbit serum; occurs with normal serum).

Wollman [335] (separated antilytic from antibacterial antibodies by special technic).

Wollman and Brutsaert [337] (lytic agent is a new antigenic entity which does not exist in the sensitive bacteria).

Wollman and Wollman [341] (the antigenic properties of principles active on germs of same group seem to be independent of their lytic action).

Bruynoghe and Dubois [68] (lytic agent can be precipitated by antilytic serum).

Flu [134] (complement-fixation involving bacteriophage and bacterial extracts).

Weiss [380] (lytic agent not destroyed by antilytic serum).

Arnold and Weiss [16a] (studies with bacterial protein-free antigens).

Gregorieff [142] (antilytic serum reduces number of areas not their size).

Wollman [334] (inactivation of bacteriophage by antilytic serum and many other points of serologic interest).

ntibacterial Serum:

Hauduroy [171] (suggested mechanism of action of antibacterial serum).

Fejgin [124] (antilytic serum action in serums prepared by three naturally resistant Shiga cultures).

Osumi [268] (tested action of serums immune to [1] lytic filtrate, [2] autolysate in distilled water and [3] normal sensitive culture. Only the first antilytic).

Machardo and da Costa Cruz [239] (action of antilytic and antibacterial serum the same).

Da Costa Cruz [100] (suggests explanation of antilytic action of antibacterial serum).

Da Costa Cruz [98, 100] (in strong antibacterial serum the bacteria reproduce without regenerating the principle: symbiosis does not exist in this case).

Kauffman [210] (two distinct antibodies observable in an antilytic serum—an antilysin and antibacterial immune body).

Kuttner [225] (failed to observe antilytic effect of antibacterial serum).

Bruynoghe and Dubois [69] (no difference in the agglutinins produced from sensitive and resistant cultures).

Wollman and Goldenberg [338] (antigenic value of bacterial substance dissolved by lytic agent greater than that of sensitive culture antigen; antibacterial serums do not have antilytic action).

Bordet and Ciuca [48] (serum immune to sensitive culture not antilytic; serum immune to lysogenic R type culture is antilytic).

McKinley [281] (serum immune to normal bacteria is antilytic).

Wollman [333, 334] (general summary of subject; confirms results of Bordet and Ciuca; antibacterial serum has no antilytic action).

Serologic Convergence and Bacteriophagic Reactions:

Hadley [167] (parallel serologic and bacteriophagic reactions in B. typhosus, Bact. pullorum and B. gallinarum).

Marcuse [245] (bacteriophagic and serologic relationships between B. coli strains and B. dysenteriae).

Burnet [78] (heterologous serologic and bacteriophagic reaction related to the presence of O and R type antigens).

Serologic Features of the Types of Lytic Agent Producing the Large or Small Plaques:

Watanabe [328] (action of antilytic serum correlated with large area principle and small area principle is specific).

Gratia [155] (specificity of action of the large area and the small area antilytic serums).

Gratia [156] (continuation of foregoing).

Hadley and Dabney [169 a] (antilytic serums for the α and β principles of B. paratyphosus B).

Miscellaneous Serologic Aspects:

McKinley [282] (serology of the lysogenic culture).

Ikoma [190] (binding of agglutinins by principle not observed).

Brutsaert [65] (agglutination of resistant forms).

Arnold and Weiss [16] (use of tryptic digestion to prepare pure bacteriophage antigens).

Weiss [331] (precipitation of proteins by lipoids in order to obtain pure bacteriophage antigens).

Wollman [334] (bacteriophage represents a "new antigenic entity" not pre-existing in the "normal" bacterial cell; serologic reactions of the resistant forms).

DISCUSSION

The Antilytic and Antibacterial Serums.—Under this heading the first point of interest is the question: Is the lytic agent antigenic? The first investigators who believed that they had demonstrated the existence of antilytic substances in serum, as a result of injecting rabbits with lytic filtrates, were Bordet and Ciuca. The following test was performed. To a tube of broth was added the following: 1 drop of coli culture, 5 drops of a mixture of equal parts of the lytic filtrate and fresh, normal rabbit serum. The culture did not grow. When, however, the normal serum was replaced by the antilytic serum, the culture grew, thus suggesting that the antilytic serum had prevented the normal action of the bacteriophage. These workers also demonstrated that the antilytic serum did not have an ill effect on the growth of B. coli. They did not show, however, that this serum did not have any effect on the type of B. coli which grew. They observed that the antilytic serum could thus neutralize about 10 times its volume of the lytic filtrate. They also showed that the process of lysis could be inhibited by serum immune to cultures that were resistant to the coli bacteriophage. When they tested the action of serum immune to the normal, sensitive strain, they found that it could not prevent the lysis of the culture. From these results the authors concluded that the bacteriophage functioned as a specific antigen. The experiments were the incentive for extensive studies. On the whole, these experiments have confirmed the earlier results, but in some cases, as will be seen, the method of experiment followed renders the conclusions open to a second interpretation.

An interesting and important aspect of the reaction was first developed through experiments reported by Bruynoghe in 1923. It appeared that the organisms which came out of the tubes after contact with the antilytic serum were no longer sensitive but resistant. It was further shown by Otto, Munter and Winkler at an earlier date that the antilytic serum could not prevent lysis if it was added to the mixture of bacteria and bacteriophage after the latter had become fixed to the cells. These results raised the question whether the immune serum affects the bacteriophage or the bacterial cells, a matter which will be discussed presently.

Regarding the fate of the bacteriophage after inhibition by the antilytic serum, Bordet and Ciuca at first believed that it was destroyed, but d'Herelle and Eliava were able to show that it was only inhibited and could again be recovered. Weiss, furthermore, in 1927, demonstrated that digestion by trypsin of the mixtures of serum and bacteriophage left the lytic agent still active.

A number of workers have believed that they could demonstrate that antibacterial serum, like the antilytic serum, was inhibitory to lysis. Machado and da Costa Cruz reported that the neutralizing effect of both was the same. D'Herelle, however, reported the opposite and showed that when an antibacterial serum was added directly to the bacteriophage the number of lytic units was not reduced, even after a period of six months. Hauduroy supported the view of Machado and da Costa Cruz, but found that more of the antibacterial serum was required to produce the same inhibiting effect. Kuttner failed to observe the antilytic effect of either antilytic or antibacterial serum on lysis. Urech and Pache showed that serums immune to sensitive dysentery cultures and to lytic filtrates all gave complement-fixing and agglutinating reactions with the homologous bacteria. The antilytic serums were weaker in action but possessed antilytic power against the homologous bacteriophage. The degree of antilytic action corresponded with the time that the bacteriophage and the serum had been in contact before adding the mixture to the bacterial culture previously spread on agar plates.

Prausnitz, after studying the neutralizing effect of antilytic Flexner serum on homologous lytic principle, reported that there was a progressive diminution in the number of lytic areas, but not in their size. Some of the corpuscles resisted the action of the serum for four days. Seiffert, however, a year later, showed that when antilytic serum in small doses was added to tubes containing broth, bacteriophage and bacterial cells the results were not to diminish the number of areas, but to reduce their size. As he increased the amount of serum, the areas became smaller and smaller until they finally disappeared.

Gregorieff, working in Wollman's laboratory in 1927, confirmed the results of Prausnitz, namely, that the neutralizing power of antilytic serum is manifested by a reduction in the number of areas. She interpreted this result as indicating unequal sensitivity on the part of the lytic corpuscles, a belief which may have some justification in view of the results obtained by Gratia at an earlier date on the complementary effect of the large area and the small area principles. She believed that this circumstance underlay the power of adaptation of the bacteriophage to unfavorable conditions, as reported by Prausnitz, Asheshov, Jansen and others.

The first studies indicating a duality of a bacteriophage active for a given species were those of Bail,[26] Bail and Watanabe[27] and of Wantanabe. Bail demonstrated the existence of the large and small area strains, arising from a pure-line, and presented his "g" and "r" types; also an intermediate form. In 1923, this line of study was extended to cover the serologic aspects by Watanabe, who dealt with antilytic serums produced by injection of filtrates representing the large and small types of lytic area. Watanabe's method was to mix the serum with the dilutions of the principle and incubate it at 37 C. before adding the mixture to the culture on solid medium. The results showed that the serums immune to these "elementary strains" of the lytic agent were highly specific for the respective types. This was confirmed by Gratia.[155] Bail observed, moreover, that the injection of sensitive bacteria alone did not produce antilytic serums, although this result was attained by the injection of resistant cultures which were also lysogenic.

Gratia, who has presented the clearest picture of the behavior of the large and small area strains of the bacteriophage, and their relation to the smooth and the rough culture types, showed in 1923 that serum produced from a "large" principle completely neutralized the corresponding principle and partly neutralized the "small" principle. When added to the original, or "mixed" (strong) bacteriophage, it inhibited the production of all of the large areas and a part of the small. Conversely, the serum immune to the "small" principle neutralized completely the corresponding principle, and partly the "large" principle. Added to the original principle ("mixed"), it inhibited the production of all of the small areas and a part of the large. It thus appeared to the author that the two principles were clearly specific, but that there existed between them a certain "parenté antigènique." Thus, as might have been expected, a serum immune to the mixed ("large" and "small") principle neutralized both strains of the bacteriophage equally.

From the observations cited previously, it may well appear that there is some variance of opinion regarding the mode of action of an antilytic serum on the bacteriophagic corpuscles, although for the moment the question of the direct or indirect action may be left unanswered. Is it to diminish their number, or is it to diminish their energy of action, and thereby to reduce the size of the areas that they produce? Taking all the data into consideration, it seems to me that the results may be interpreted, to an extent at least, in the following manner, and by the aid of the following facts.

It will be noted that the method of experiment of Seiffert differed from that of Prausnitz and of Gregorieff. The former brought his immune serum into contact with the bacteriophage and cells combined, while the latter brought the immune serum into contact with the bacteriophage alone, then added the mixture to the cells. It was seen that, in the former case, the size of the areas was reduced; in the latter case, the number of areas. In Seiffert's case the antilytic serum had considerable opportunity to influence the cells; or, at least as great opportunity as it had to influence the lytic corpuscles. In the experiments of Prausnitz and of Gregorieff, as I understand their methods, the antilytic serum had a greater opportunity to influence the bacteriophage, since the latter was treated separately before the mixture was added to the cells. One modification of the test (for the purpose of control), which does not seem to have been carried out by any of the investigators mentioned, would be to add the antilytic serum to the cells and incubate the mixture, and then to add the bacteriophage.

In any case, the question arises: What was the influence of the antilytic serum on the culture in Seiffert's case? From much work on the serologic aspects of microbic dissociation it is known that the contact of sensitive S type cells with homologous immune serum is followed by transformation toward the R type culture, which is more resistant to α type bacteriophage action. One knows, moreover, from the studies of Gratia, that the action of β lytic agent on the R type culture is characterized by the production of small plaques. We know, further, that the transformation of the S type culture to the R under the influence of immune serum is gradual and progressive. With these considerations in mind it seems natural that, when the antilytic serum (which seems to possess the equivalence of an anti S-type serum) is added to a mixture of culture and bacteriophage, the serum should so influence the culture, by initiating the S to R transformation, that more of the small areas would be produced; and that, as Seiffert observed, the increase in the amount of the antilytic serum (thereby forcing more rapidly the transformation) would be followed by the production of smaller areas. This view receives support from the observation of Bruynoghe and Wagemanns, already mentioned, that the organisms recovered from a tube to which antilytic serum has been added and in which lysis had failed to occur, were resistant to the principle concerned. It seems probable that, in this case, the sensitive organisms had been transformed to, or toward, the R type under the serum influence; and that the transformation occurred before the lytic agent was able to operate on the original cells of the culture. This view of the matter also receives support from the observations of Otto, Munter and Winkler that, if the bacteriophage is added to the sensitive cells about an hour before the antilytic serum is added, the progress of lysis is not interrupted by the serum influence.

The foregoing interpretation does not militate against the view that such a thing as an antilytic serum may exist. It is presented merely for the purpose of indicating that, in certain experiments reported, there are grounds for a double interpretation, and that the transforming action of an immune serum on the culture of the substratum must be ruled out before we can justifiably conclude that the immune serum action is actually antilytic. Fortunately, in the literature, there are some instances in which the tests of the antilytic serums were controlled by correlated experiments including observations on the influence of serums immune to the normal, sensitive cultures.

Osumi, in 1924, studied a group of serums immune (1) to colon lytic filtrates, (2) to sensitive coli culture and (3) to autolysates of coli culture in distilled water. The antilytic serum, when added to the lytic agent and then tested against the culture, showed a strong inhibitory effect, while the other serums did not show any. This would seem again to indicate the existence of a distinct antigenic function of the lytic agent. Indeed, the only possible objection to such an interpretation is based on the observation by Wollman and Goldberg that serums immune to dissolved (by the bacteriophage) bacterial substances are much more active than serums immune to antigens made from bacterial cultures of the sensitive type. And, so far as we know, the transforming action of such serums on the sensitive culture might be correspondingly greater. Of this fact, however, there is not any actual evidence so far as I am aware.

Wollman and Brutsaert, in 1925, again took up the study of the action of antilytic serums and attacked the problem with a somewhat new technic. They tried to relate the bacteriophagic antigen to antigenic constituents of the microbes at whose expense the lytic agent was developed. They argued that since antilytic serums seem to inhibit lytic action the agent must, therefore, possess antigenic function and could be neutralized by the antibody produced. In such a case, in which the agent may concern a product of the bacteria (pro-ferment or ferment), one ought to discover this antigen in the bacteria themselves. In other words, serums prepared with the bacteria, intact or lysed artificially, ought to inhibit the action of the bacteriophage, at least to a degree. Without going into the details of their experimental method, devised to meet this question, the conclusions of these authors may be briefly noted—namely, that since the serum immune to the bacteria themselves seemed not to embrace antilytic powers, while these were manifested by the antilytic serums, the bacteriophage must be regarded as constituting a new antigenic entity, not found in normal bacteria. Although this may be true, here it may be said again that the experimental results of these authors do not absolutely preclude another interpretation, based on the circumstance that the knowledge of the serologic aspects of microbic dissociation convinces one that there is no necessary symmetrical serologic relationship between different cyclostages of any single, pure-line culture. If the bacteriophage units were stages in the development of the culture itself, there is not any reason for anticipating that they would show symmetrical antigenic relationships with the mother culture. Indeed, the opposite would be expected. This subject I have considered in detail, with adequate evidence, in a former publication.

To the foregoing citations it may be added that Kasarnowsky and Timokin-Schukoff, in 1925, reported the results of experiments believed by them to demonstrate that the bacteriophage is a specific antigen, and is composed of bacterial substance—an important view, as will be seen later.

In 1926 da Costa Cruz again took up the study of the antibacterial serum as opposed to the antilytic, and concluded that these differed only in the degree of action; that the action of both was due to the bacterial antibodies. Later he presented the results of other experiments showing more conclusively, as he believed, that the antibacterial serum inhibited the regeneration of bacteriophage, whatever the time of contact with the lytic agent and the bacterial cells might be. He believed that the fact that the bacteria could reproduce in such mixtures without regenerating the bacteriophage proved that a symbiosis between the lytic agent and the bacteria did not exist under these conditions.

Referring to the conclusions of da Costa Cruz, Wollman in 1927 believed that heavy doses of antibacterial serum might interfere with the progress of lysis, but that in such a case this was because it was strong enough to produce an agglutination of the bacteria. He thus reiterated that only antibacteriophagic serums were antilytic and championed Bordet and Ciuca's view of the matter. There was no community of antigenic structure between the bacteriophage and its homologous culture. As we shall see, however, this eventuality does not tell us anything of significance regarding the possible biological relationship between the culture and its bacteriophage.

From the preceding exposition it might appear that the serologic studies on the antigenic nature of the bacteriophage would be greatly aided if the agent could be freed from the products of lysis and of bacterial metabolism with which it is, by necessity, commonly associated. Some attempts to accomplish this have been made.

Arnold and Weiss attempted in 1925 to surmount the difficulties by tryptic digestion of the bacterial residues. They found that the resulting fluid contained no bacterial proteins, and that the injections of such fluids into rabbits resulted in the production of serum possessing only neutralizing antibodies, or antilysins. No complement-fixing or bacterial antibodies were present. Similar results were obtained by Weiss, at a later date, as a result of precipitating the foreign protein in cultures of bacteriophage by means of lipoids.

In concluding these references to the antilytic serums, mention may be made of d'Herelle's conception of the fundamental unity of all bacteriophage strains. If this were true, and the bacteriophage were really antigenic, one would expect that this unity would be manifested by complement-fixation tests. D'Herelle and Eliava at first believed that they could so demonstrate this unity. Their tests seemed to reveal the fact that any antilytic serum contained sensitizers which caused the fixation of complement in the presence of the lytic filtrate of any other race of bacteriophage. D'Herelle was able to show later, however, that fixation also occurred when he used as antigen, instead of the lytic filtrates, the autolysates of any bacterial species; therefore, he came to the conclusion that the bacterium contained the common antigen. His latest conclusion is that, if any sensitizers for the lytic agent are formed, they have but little potency. To this it should be added that Sanderson [298] in 1925 believed that he had demonstrated that the actual explanation underlying d'Herelle's earlier results rested on the following circumstance: that nonspecific fixation occurred with rabbit serum. He found the same degree of fixation occurred with normal rabbit serum. Arnold and Weiss, moreover, demonstrated that antilytic serums prepared from lytic filtrates from which the bacterial proteins had been removed by tryptic digestion, possessed no complement-binding antibodies—only the antilysins. These results, also supported by the later work of Weiss,[331] seem to uphold the conclusions reached by Sanderson.

Regarding the reciprocal relationships between antilytic serums for B. typhosus and for B. dysenteriae Shiga, Wollman and Wollman [341] have recently presented data of interest. First, they observed that, when Shiga bacteriophage was brought into contact with Shiga culture on agar, it determined almost complete lysis; when added to typhoid culture, it gave areas 0.5 to 1.0 mm. in diameter. Under similar conditions a typhoid bacteriophage gave complete lysis of typhoid culture and only a few areas 4 to 5 mm. in diameter on Shiga culture; these areas resembled those on typhoid itself. Each bacteriophage thus acted as a "feeble" principle toward the culture of the other species. If the typhoid principle was passed several times through Shiga culture, the character of the plaques remained constant although their number increased. At the same time the "virulence" of the principle was diminished for typhoid. As the Wollmans well pointed out, these results are little in agreement with Bordet's conception of "concentration." If one admits that the Shiga culture is less sensitive to the typhoid principle than the homologous culture, and that the more and more pronounced lysis obtained on Shiga may be due to an increase in concentration of the agent, one can scarcely believe that, at this concentration, the bacteriophage would become much less active for the typhoid culture, in which it caused, at the beginning, a total lysis. There is, thus, a modification of the virulence of the principle—according to the Wollmans, a true adaptation to the new culture, with partial loss of activity for the old.   These results seem dependent on the number of lytic corpuscles, not on the number of sensitive bacteria.

From the serologic tests accompanying these experiments it appears that serum immune to the typhoid principle neutralized not only the action of this principle on typhoid and Shiga cultures, but the action of the Shiga principle on both of the species.   Moreover, while the typhoid principle was inactive against both dysentery Y and Flexner cultures, the serum prepared with the principle neutralized perfectly the lytic action of the Shiga bacteriophage for these two species.   From these experiments the Wollmans concluded that the antigenic properties of bacteriophages active against members of the same group appeared to be independent of their lytic action.

From the somewhat confusing array of reports which have now been examined, what shall we conclude regarding the bacteriophage as an antigen?   I do not believe a final answer can be given at present. All we can safely conclude is that the greater body of evidence favors the view of its antigenic nature, but at the same time seems to demonstrate that evidences of the antigenic power are often concealed by a second reaction-factor which has a blanketing effect.   This last is the circumstance that the immune serum (antilytic or antibacterial), as employed in many of the experiments reported, tends to produce, per se, a modification of the sensitive culture toward the more resistant R type, in which it may be less susceptible to the influence of the bacteriophage. The experiments which furnish the clearest evidence of antilytic action of bacteriophage immune serum are those in which the serum was incubated with the lytic filtrate before it was added to the sensitive cells; here, a direct influence seems sometimes to have been exerted.   It seems to me, however, that the conclusions on this matter can never have a secure support until methods are devised to obtain the bacteriophage in a relatively pure state, at least freed from the bacterial protein and metabolic products, before it is used as an immunizing antigen; or, it might be added, until a method has been found to remove completely from the assumedly antilytic serum the bacterial antibodies, as attempted by Otto and Winkler.   In view of the diverse antigenic components which, as is now known, make up the average bacterial culture, success by the latter method is extremely improbable.   There remain, however, such methods as those employed by Arnold and Weiss, or by Weiss, involving tryptic digestion or precipitation of the bacterial residues; or perhaps some means may be found to utilize the fact reported by de Poorter and Maisin and by the Clarks,[89] namely, that the bacteriophage is quickly and effectively absorbed by certain charcoals, particularly (Clark) by that known in the chemical warfare service as "Rankanite A." The employment of some such method might make possible the production of strictly antilytic serums through whose use the masking reactions attributable to the influence of bacterial antibodies would be completely ruled out. From this point of view, the experiments of Arnold and Weiss and of Weiss are of much significance, and their confirmation and extension are matters of considerable importance.

It now becomes apparent from the studies of Bail, Watanabe, of Gratia, Kline and of Dabney and myself that one can no longer regard the bacteriophagic antigen as if represented by homologous units. It is not homogeneous, but possesses a dual nature, each component of which (the α or the β units) seems to serve as an independent antigen. The apparently antilytic serums, derived respectively from the α and β units, do not appear to be symmetrical in their action. This must be taken into consideration in future work in this field. From now on it will scarcely be sufficient to speak of "an antilytic serum." The worker in question must know whether he is dealing with an anti-α or an anti-β type serum; also whether the cyclostage of the culture of the substratum represents the "normal" sensitive (S), the first resistant (R) or the second resistant (R^2) type; in other words, whether the substratum is suitable for the generation of the α or the β bacteriophagic units, or both. It seems to me that a consideration of these aspects of the bacteriophage reaction will eventually make possible a reconciliation of many of the divergent results previously reported.

Serology of the Sensitive and the Resistant Types.—The serologic aspects of the bacteriophage reaction naturally include the serology of the various culture types concerned in the reaction. Although several among these possess much significance in the lytic reaction, the present knowledge of them is almost exclusively limited to the beginning and the end stages, namely, the normal sensitive (S) and the secondary resistant forms. Lying between these are intermediate culture types, undoubtedly of great significance in the reaction, but of which there is, as yet, little knowledge; here, the only fact that is clear is that these forms closely resemble the intermediate types observed in the normal dissociative reaction. My present considerations will therefore be limited to the sensitive culture and the secondary resistants, of which, as already noted there may be at least three different types.

In 1921 Arkwright [8] first drew attention to the two distinct forms of culture, S and R, which belong in the normal dissociation series. In 1922, Bordet [36a] showed that he could obtain a "strong" lytic principle which was able to effect the lysis of both forms, and also a "weak" principle which brought about the lysis of the S culture only. In addition, he pointed out that when the weak principle worked on the purified S type culture it transformed this culture within a few hours into a type which he termed "P," but which seemed to be analogous to Arkwright's R form. It was from such experiments that Bordet conceived the notion that the bacteriophage played a rôle in directing the evolution of the culture, and was a controlling factor in mutations.

The question arose in the mind of McKinley [252] as to the possible existence of serologic differences between the sensitive and resistant cultures (Bordet's B and P types), such, perhaps, as those differences which Arkwright had shown to exist both serologically and in other respects between his S and R types. Serums were prepared against each type. In brief, McKinley's results were as follows: When tested, each agglutinated strongly only its homologous culture. Therefore, McKinley rightly concluded that the weak bacteriophage had transformed the antigenic specificity of the B (S) type culture.

In 1923, Fejgin [124] obtained three modified strains of a Shiga culture which appeared spontaneously, gave agglutinative growth in broth and were resistant to the action of a Shiga bacteriophage. The serums of rabbits immunized with these cultures were all antilytic. These cultures were manifestly the R types from normal dissociation. This is the only instance of which I am aware in which R type immune serum has been tested for its antilytic power. As previously pointed out, how to interpret these results is still a question.

Next we may consider the nature of a serum produced by the injection of a resistant culture of the lysogenic type. Such a type was first reported for B. coli by Lisbonne and Carrère. [236] They showed that this culture, although remaining unlysed, was able to stimulate the transmissible lysis in Shiga cultures. McKinley [251] undertook to ascertain whether serum immune to this culture was able to neutralize its lytic capabilities. The serum obtained from rabbits possessed high agglutinating and precipitating power; it also inhibited the lytic action on Shiga cultures possessed by the lysogenic strain. McKinley concluded that one can obtain an antilytic serum from "normal cultures" only when these are of the lysogenic (Lg) type. It appears from this study, however, that the antilytic serum only neutralized the principle that was already formed; it did not destroy the lysogenic ability of the Lg culture.

Relation Between Serologic and Bacteriophagic Reactions among Related Bacterial Species.—In 1926, I [187] called attention, in a brief publication, to the observed parallelism between bacteriophagic and serologic response in B. typhosus, Bact. pullorum and B. gallinarum. Here the inter-reactions were practically indicated. On the strength of these observations I then raised the question whether, in general, the range of bacteriophagic reaction of a given strain showed a parallelism with the range of serologic affinities of the same bacterial species; and particularly whether the range of bacteriophagic reaction was in any way correlated with the range of serologic reactions to be observed in connection with the O or R types. As a result of the experiments of Schutze [302] on intestinal organisms, of Goyle [145] on B. typhosus, B. enteritidis and B. dysenteriae and of Balteanu [352] on the cholera vibrio, it is known that the "serological cosmopolitanism" or serologic convergence of Schutze [302] is largely dependent on the common possession of the heat-stable antigens characteristic of the O and R forms of culture. It would be of much interest to ascertain the extent to which, if at all, bacteriophagic reactions may be dependent on the same common antigens in the bacterial cells of the substratum.

An investigation bearing on the point at issue was reported by Marcuse [246] in 1926. This investigator, although not studying the S, O or R forms of culture particularly, was able to demonstrate a relation between serologic and bacteriophagic reactions involving dysentery Y and certain coliform strains. The lytic agent for the former culture was active for the latter cultures. The coliform strains, moreover, possessed the ability to absorb a considerable amount of the agglutinin from a serum immune to dysentery Y. Coliform strains that were insensitive to the Y bacteriophage did not absorb such agglutinins, or absorbed them to a much smaller extent.

But the clearest picture of the parallelism between serologic and bacteriophagic reactions in different bacterial species is found in the recent study of F. M. Burnet,[78] in which he not only confirmed, but also extended in considerable degree, the facts and the interpretation of my earlier, but brief, experiments. Burnet studied the serologic and bacteriophagic reactions of the sensitive and resistant (rough R) forms of B. typhosus, B. enteritidis and Bact. pullorum.

His work dealt especially with the possible nature of the specific antigenic constituents of the cultures that might be responsible for the concurrence of the observed reactions. He found that, among Salmonella types, those strains possessing the same heat-stable agglutinogens as B. enteritidis reacted similarly to a series of lytic agents that had been developed on enteritidis strains. "The strains with no O agglutinogens in common showed no susceptibility to the enteritidis phages—with one exception." "The R forms of all Salmonella types show much the same behavior toward phage—a fact that can be correlated with their 'serological cosmopolitanism' (Schutze)." Strains of lytic agent developed on rough enteritidis cultures "are active against rough variants of most Salmonella types, the range corresponding to the parallel extent of common, heat-stable agglutinins amongst such forms." Burnet thus observed a coordination between agglutinin-absorbing capacity of cells and their "bacteriophage-absorbing" capacity, and believed that both were due to the presence of the heat-stable antigen. He regarded his results as a whole, pointing to the biologic independence of the bacteriophage particles. I believe that Burnet's results and conclusions are of great significance in that they furnish grounds for suspecting that, among the intestinal forms at least, the bacteriophagic reactions tend to follow the lines of antigenic relationship revealed by the common antigens of the R type (and perhaps the O type) cultures.

Conclusions.—Although it is admittedly difficult to grasp the significance of the often contradictory results of serologic studies dealing with the bacteriophage and its reactions,* we may at least accept certain probabilities. First, the bacteriophage is antigenic, or weakly so. Whether antilytic power can be demonstrated unequivocally in a serum immune to lytic filtrates seems to depend on the experimental method employed; for it cannot be doubted that antilytic power of a serum may easily be masked by a failure of lysis due, in reality, to the transforming influence of an antibacterial serum applied to the culture of the substratum. On the other hand, the antilytic reactions of pureline "elementary strains" of bacteriophage seem to be to a considerable degree specific; moreover, there seems to exist an antigenic diversity

* It will be apparent to the reader that our knowledge of this aspect of the bacteriophage problem will not be furthered materially until there have been performed experiments which provide not only for the purity of the bacteriophage used (that is, freedom from extraneous organic substances), but also for a pureline principle (that is, a principle constituted exclusively of either the α or the β units). Up to the present time nearly all the tests reported have been performed by the use of "a bacteriophage," without specification as to whether an α, a β or a "mixed" principle was employed.

between the "large" and the "small" strains of bacteriophage, representing a phenomenon for which an explanation is not yet available.

It appears further that the only "pure" antilytic serums obtained up to the present contain neither complement-binding, agglutinating nor precipitating, but only antilytic, antibodies, if one may accept the conclusions of Arnold and Weiss and the more recent work of Weiss on lipoidal precipitation. The absence of these antibodies in the antilytic serum does not, however, militate against the view that the bacteriophage may be a living agent, or even a living element related to the species of the substratum, for the knowledge of the dissociative reactions makes it clear that certain cyclostages of common bacterial species exist in whose immune serums one or another of the expected, parental antigens is not found.

Since the antilytic serum manifestly does not destroy the bacteriophage, or effect any permanent neutralization, it is impossible for the present to conclude what the action of the antibodies on the corpuscles actually is.

Of all the work thus far accomplished on the serologic aspects of the bacteriophagic reaction, I regard as most significant, and pregnant with future possibilities, those observations of F. M. Burnet, dealing with the parallelism between the serologic and bacteriophagic reactions; and especially those dealing with bacteria possessing the common heat-stable O or R antigens. These results, taken in conjunction with the observations of Gratia [155, 156] (supporting the correlation of heat stability or heat lability of pure-line bacteriophage strains with the heat-stable or heat-labile characters of the antigens dominating the culture types in which these strains of bacteriophage, respectively, arise), add to the evidence that the bacteriophagic units are in some way biologically and genetically related to the cells in which they are generated.

8. THE RÔLE OF THE BACTERIOPHAGE IN DISEASE AND IN IMMUNITY

In many of the writings of d'Herelle the relation of the bacteriophage to disease has been a conspicuous and interesting feature. The references have largely centered about (1) its relation to immunity or susceptibility , (2) its therapeutic value or (3) its prophylactic value. Regarding the first point, he has expressed the view that the inception of certain communicable diseases, such as typhoid and dysentery, is dependent on the absence from the body of a strain of bacteriophage capable of dealing with the invading parasite; or on the fact that the parasite becomes resistant to the lytic agent already present and active. In prophylaxis and therapeutics he lays stress on the valuable results following the administration of lytic filtrates, used in various ways. More recently, some investigators have supported d'Herelle's claims, and even furthered the evidence in favor of them. A smaller number have not seen in the results of bacteriophage administration a tendency toward the suppression or modification of the disease course. Some have even observed an increase in susceptibility in animals as the result of treatments with bacteriophage which were not carefully given. As will be observed, most of the tests of therapeutic value have been concerned with human infections—only a few with the natural infectious diseases of the lower animals, in which the experimental conditions are, on the whole, much more favorable, and in which d'Herelle achieved his greatest successes.

REFERENCES

Relation of Presence of Bacteriophage to Presence or Severity of the Disease:
  D'Herelle [192] (fowl typhoid, barbone, typhoid, dysentery, suppurative infections).
  Ciuca and Manoliu [88] (relation not established in typhoid).
  Wagemans [327] (failed to observe relation to incidence of typhoid).
  Boulet [49] (as Wagemans).
  Larkum [227, 228] (relation to incidence in urinary infections).
  Stolz [319] (relation to prognosis in typhoid fever).
Behavior of Bacteriophage After Injection Into Body:
  Krestownikowa and Gubin [219] (appears in five minutes in all organs; disappears after 6½ to 8½ hours).
Bacteriophage in Typhoid Infections:
  Beckerish and Hauduroy [28] (first report; favorable results; same in paratyphoid infections).
  Hauduroy [174] (favorable results; double rôle of lytic filtrates).
  Richet, Aserad and Delarve [284] (favorable results).
  Smith [313] (favorable except in typhoid bacteremia).
  Hauduroy [179, 180] (favorable results; positive blood cultures became negative).
  Lisbonne and Boulet [233 a] (favorable results).
  Gjorup [189] (unable to modify the carrier state).
  Otto and Winkler [272] (no favorable effect).
  D'Herelle [198] (favorable results).
  Stolz [319] (relation of bacteriophage to agglutination in typhoid).
  Hauduroy [181] (the influence of bile).
Bacteriophage in Dysentery Infections:
  D'Herelle [198] (favorable results in treatment).
  Otto and Munter [269] (results not favorable).
  Davison [104] (results not favorable).
  Euguchi [119] (results not favorable).
  Peieira [274] (results very favorable).
  Arnold and Weiss [14 a] (favorable in rabbits).
  Da Costa Cruz [98] (gives results surpassing all other methods).
  Spense and McKinley [314] (results very favorable).
  Lesbre [229] (results less favorable than from use of antitoxin).

Bacteriophage in B. Coli Infections (Mainly Urinary):

Beckerish and Hauduroy [28 a] (results very favorable in pelocystitis).

Hauduroy [182] (results very favorable in pyelonephritis).

D'Herelle [193] (results very favorable in various infections).

Courcoux [102] (results favorable in urinary infections).

Philibert [278] (same).

Marcuse [243] (favorable in experimental infection in rabbit).

Alphonsi [4] (favorable results in pyelonephritis).

Zdansky [363] (method of obtaining principle from sewage for use in urinary infections).

Cowie [316] (favorable results in pyelitis).

Delsase [105] (favorable results in urinary infections).

Wollman [355] (limited protection in guinea-pig infection).

Larkum [227] (favorable effects in pyelitis in children and adults).

Caldwell [81] (sewage as source of coli bacteriophage for urinary organisms; favorable results in treatment).

Bacteriophage in Staphylococcus Infections:

Bruynoghe and Maisin [72] (first favorable result in furunculosis).

Gratia [151, 152] (favorable results in folliculitis, furunculosis, subcutaneous abscesses, etc.; polyvalent principle).

Barbosa (favorable results in staphylococcus cystitis).

McKinley [250] (favorable results in suppurating wounds).

Gougerot and Peyre [146] (favorable results in pustular sycosis).

Hauduroy [175] (sensitization of young rabbits by principle).

Bazy [388] (favorable results in surgical infections).

Hauduroy [189] (summary of recent work in therapeutic use of bacteriophage).

Delsase [105] (favorable results in urinary infections).

Bacteriophage in Streptococcus Infections:

McKinley [251] (favorable results).*

Dutton [110, 111] (favorable results).*

Bacteriophage in Other Human Infections:

Zoeller and Manoussakis [369] (results not favorable in B. pyocyaneus infections).

Bruynoghe and Maisin [72] (favorable in anthrax infection).

D'Herelle [191] (results favorable in plague).

Beckerish and Hauduroy [28] (favorable in paratyphoid fever).

Pelouse and Schofield [273] (of some value in gonorrhea).*

Sonnenschein, K. [313 a] (results favorable in chronic atrophic rhinitis).

Bacteriophage in Natural Infectious Diseases of Animals:

D'Herelle [188] (favorable results in fowl typhoid).

D'Herelle [188] (favorable results in barbone).

Louet [237] (confirmed d'Herelle's results on barbone).

Pyle [291] (results unfavorable in Bact. pullorum infections).

Levy [232] (mouse typhoid; could not modify course of infection).

Topley [331] (same; favorable results from intraperitoneal administration).

Richet and Hauduroy [296] (same; results not favorable).

Bronfenbrenner and Korb [67] (same; results not favorable).

Contraindications, Precautions, Methods; General Considerations:

Da Costa Cruz [92] (methods and amounts; ampules furnished by institute).

---

* The study here reported probably did not concern the bacteriophage in the d'Herelle sense. The results suggest only a strong dissociative action of filtrate for the homologous culture. In later work Dutton probably worked with actual streptococcus bacteriophage.

D'Herelle [193] (plague, method and amounts).

Gratia [132] (method and amounts in staphylococcus infections).

Larkum [228] (pyelitis, methods and amounts).

Zdansky [845] (method of obtaining principle for use in pyelitis and urinary infections).

Arnold and Weiss [16a] (rate and degree of opsonin production from injections of [1] bacteriophage, [2] living typhoid bacteria and [3] killed typhoid bacteria; toxicity of typhoid and dysentery bacteriophage).

DISCUSSION

Therapeutic and Prophylactic Uses: Typhoid Fever.—In diseases produced by the members of the colon-typhoid-dysentery group the results of treatments with bacteriophage have, in general, been favorable; at least the favorable reports outnumber the unfavorable ones. Since the effectiveness of the method was first announced by Beckerish and Hauduroy for typhoid infections in 1922, many confirmations have appeared. Although Smith reported failures in patients who gave positive blood cultures, Hauduroy found that in his cases positive blood cultures soon became negative. In regard to typhoid carriers, who, it is known, often carry organisms of the resistant R type, Gjorup found that bacteriophage therapy was without effect. In explaining the unfavorable instances it must be borne in mind that the site of carrier infection is often in the gallbladder, in which the presence of bile might militate against favorable bacteriophage action, as suggested by the work of Hauduroy; [181] moreover, that the typhoid group is highly heterologous with respect to the lytic influence of a single strain, as d'Herelle early pointed out. In certain experiments which will be reported in detail at a later date, I have found that, in vitro, a single typhoid principle seldom is effective in producing inhibition in more than one half of the heterologous strains on which it is tested. This implies the desirability of employing a highly polyvalent filtrate, but I am not aware that the use of such has been reported in the treatment of typhoid cases. I have prepared such polyvalent filtrates from ten or more strains of B. typhosus, and found them active in vitro on all the individual strains concerned, and on Shiga. I have not had opportunity, however, to use this polyvalent filtrate in clinical cases of typhoid fever.

With reference to the presence of the bacteriophage in the blood of patients with typhoid, Hauduroy observed it in a large number of cases. Stolz noted that the presence in the blood did not interfere with agglutination (in vitro) occurring within two hours. He believed that if the flocks became dissolved within two hours the bacteriophage for the typhoid bacillus was present and the prognosis in the case was good.

Dysentery.—Since the first early reports of d'Herelle, favorable results have also been obtained from the therapeutic use of the bacteriophage in dysentery. The results of Pereira and of da Costa Cruz. in particular, support the belief in its prophylactic value. The reasons for Davison's and Euguchi's failures are difficult to understand. In Brazil, the Institute Oswaldo Cruz has undertaken the dissemination of the dysentery bacteriophage, in ampule form, over a wide area. A similar extensive use is now being made in Egypt under d'Herelle's supervision. It may be added that the difficulty that might be expected from the Shiga toxin seems to be of small consequence if the filtrates are properly "ripened" for a sufficient time.

B. coli Infections.—The first investigators to employ the bacteriophage in B. coli infections were Beckerish and Hauduroy. Since that time many have confirmed its value. Indeed, it may be said at present that the treatment of urinary infections appears to be one of its most significant fields of use. In pyelitis, pyelonephritis, cystitis and related infections the results seem to be equally valuable. The most recent reports include those of Cowie and of Larkum. Cowie has experienced much success in pyelitis. His methods and results of administration, moreover, have been such as to afford grounds for observations on the possibility of an antibacteriophagic immunity in his patients. Indications of this were never seen. Larkum's results are of special interest, since they concern themselves with some of the more fundamental aspects of the subject, such as the occurrence, source, permanence and fate of the bacteriophage in the urine, and the character of the colon cultures occurring in the urinary infections. Larkum found the lytic agent present in about 36 per cent of patients in whom some type of urinary infection was present. Urines which did not contain bacteria were free from the bacteriophage, while more than 40% of infected patients carried a strain of B. coli capable of lysis by some strain of bacteriophage. Only in exceptional instances did chronic cases yield bacteria other than the resistant strains of B. coli, acute cases seldom showing the resistant type. It is a point of interest that males, with one exception, were never a source of bacteriophage or of susceptible bacteria. Four patients (including both children and adults suffering from pyelitis) who received bacteriophage (kidney, bladder and subcutaneous injection) showed definite improvement. The author conservatively leaves open the question as to the part played by the bacteriophage itself in these cases. A second part of this valuable study involves the conditions surrounding coli infections and bacteriophage in rabbits.

Caldwell,[81] in her study of coli strains from urinary infections, reported that 95% were susceptible to lysis by bacteriophage obtained from sewage. Zdansky [345] had presented similar results.

Staphylococcus Infections.—In staphylococcus infections the value of bacteriophage administration first reported by Bruynoghe and Maisin has been observed by many investigators and physicians. In this field of therapeutics Gratia has made many of the most important contributions, and a large number of investigators have made use of his polyvalent bacteriophage, "strain H." This was employed by Bazy in his important series of investigations on the value of the bacteriophage in surgical infections of great variety. In some cases he brought the agent into direct contact with the lesions, while in others he injected it subcutaneously some distance away. He concluded that in either case the treatment caused a rapid regression of the infective process and of the lesions, and effected the restitutio ad integrum within the shortest possible time. It was also noted that the staphylococcus agent had a beneficial effect on streptococcus lesions, as had been observed earlier by Gratia. Bazy believed that the general results of this treatment did not differ greatly from those obtained by use of such bacterial extracts as the "endococcines." In this connection it may be noted that Gratia obtained favorable results by the use of lysates of cultures heated at 70 C., also by the use of lysates produced by the autolysis of cultures in sealed tubes. In view of these and related observations, Bazy was undoubtedly right in concluding that the favorable results obtained by the use of the lytic filtrates cannot be entirely explained by the action of the bacteriophage itself. The soluble bacterial products undoubtedly play an important part, and in this connection it is of interest to consider the relation of these reactions to those observed by Besredka and others, arising from the administration of the "antivirus."

Streptococcus Infections.—In streptococcus infections the value of the streptococcus lytic agent has been supported by McKinley and by Dutton. In neither case, however, is it clear that a bacteriophage in the d'Herelle sense was obtained or employed, although d'Herelle accepts this as having been the case in McKinley's reports. In Dutton's cases filtrates of cultures which manifestly showed a marked dissociational activity were employed therapeutically. The filtrates showed some degree of inhibiting power in broth, and had the ability to provoke dissociation in the sensitive cultures. They did not, however, manifest clearly other phenomena usually regarded as characteristic of the lytic agent, such, for example, as the lytic areas. Dutton admits that the reaction is not perfectly characteristic of the usual bacteriophagic action.* I believe one is safe in concluding, however, that it is a closely related reaction, and that it may represent the bacteriophage reaction itself partially obscured by some still unrecognized factors. Here, again, it is of interest to compare Dutton's and McKinley's results and methods with those of Besredka. Both Larkum [351] and Pryor [351] believe from their experience that staphylococcus bacteriophage has a favorable influence on streptococcus infections.

Other Human Infections.—Among the attempts to employ the bacteriophage in the therapeutics of other human diseases, the most recent success of d'Herelle in the treatment of plague in Egypt is of great interest, although there were only a few cases. Here, the injection of bacteriophage for B. pestis, either intravenously or directly into the buboes, appeared to be accompanied by highly favorable results. Pelouse and Schofield report some favorable effects of bacteriophage treatment in gonorrheal infections. Sonnenschein has reported favorable results in rhinitis.

Diseases of Animals.—In diseases of animals, such as fowl typhoid of birds and hemorrhagic septicemia of the East Indian buffaloes, it has appeared from the early work of d'Herelle that injections of the bacteriophage are of much value, either in combating the existent infection (fowl typhoid) or in prevention (barbone). In such natural infections as mouse typhoid, caused by B. typhi murium, the results of bacteriophage treatments, either in preventing or curing the malady, do not appear to have been so favorable. In view of the results obtained by d'Herelle in fowl typhoid, one might have anticipated similar effects in the related infection produced by Bact. pullorum, but according to the recent report of Pyle this is not the case.

Mechanics of the Protective Reaction.—Unless one employs some method to rid the lytic filtrate of the products of bacterial metabolism and of bacterial residues, the filtrate will manifestly contain at least two, and perhaps more, active agents—the bacteriophage itself and the products of bacterial autolysis, or residues of some sort. In addi-

---

* In Dutton's later cases a bacteriophage was probably present.

tion, there may be products of bacterial metabolism given off into the medium before lysis occurs. It can scarcely be doubted that each of these components of the so-called lytic filtrate plays a part of some sort; and we are therefore confronted with the important problem of evaluating the actual immunizing or neutralizing power, for the invading bacteria or their products, of these three possible constituents of any lytic filtrate.

In such work as d'Herelle conducted on fowl typhoid, it seems probable that the more or less immediate effects produced are due to the action of the bacteriophage itself rather than to the products of lysis of the bacteria. The fairly immediate result in typhoid, characterized by sudoral crisis and defervescence within a few hours, likewise suggests an immediate effect which can be explained most readily on the grounds of direct bacteriophagic action. In the immunization of the buffaloes with the principle of barbone another sort of phenomenon appears—a much delayed appearance of the immune state, seldom under four days after administration of the minimum dose. Here it seems clear that the immune reaction is attributable to the slower generation of bacteriotropic antibodies. The serum gains marked opsonizing power, as is also often observed after administration in typhoid (d'Herelle [192]).

In the reaction to bacteriophage administration in suppurative infections the mode of action of the filtrates is difficult to ascertain owing to the similar improvement, in the course of infection, which often follows the use of heated filtrates or autolysates of a different origin (Bazy, Gratia). It is true that the favorable reaction appears quickly, but here it must be borne in mind that apparently similar early reactions appear after the use of the culture filtrates of Besredka, whether administered by injection, oral route or bandaging. Indeed, there is some reason to believe that some of the favorable results of bacteriophage therapy depend on factors and immune processes not far different from those underlying the response to Besredka's antivirus.

Finally, the question arises as to the manner in which the bacteriophage exerts its influence in those cases in which its action appears to be direct and immediate. Is it because the infecting organisms in the body are actually "destroyed" by lysis, as many investigators believe, or is another factor involved? Although direct destruction by lysis may sometimes occur, especially following oral or rectal administration of the lytic filtrate, I believe we may easily attach to this reaction too great a significance, just as too great a significance has undoubtedly been attached to lysis in vitro, as compared with other accessory phenomena accompanying bacteriophage action on the culture. Just as, in the tube, the fundamental reaction determined by the bacteriophage is the enforcement of microbic dissociation, so, I believe, the most significant reaction occurring in the body of the animal is not the lysis or dissolution of the invading microbes, but their dissociation into harmless, or at least less harmful, forms. Studies on dissociative reactions of bacteria have surely taught that, in many bacterial species at least, it is only, or mainly, the smooth, sensitive S type of culture that is dangerous, since it, chiefly, harbors virulence and perhaps toxicity. Such S type organisms are commonly nonphagocytable, as has been demonstrated by many observers. We know, further, that under the dissociating influence of a bacteriophage of proper strength the virulent smooth type is transformed into the rough or other forms of resistant culture which are most commonly nonvirulent and easily phagocytosed; in other cases, it seems to disappear without leaving any microscopic cell forms. It therefore seems to me that, whenever we are justified in concluding that the chief action of the bacteriophage is direct, the actual mechanism of protection lies in its ability to determine the dissociation in vivo of the invading germs; and this means the ability to transform them into a new cyclostage (the R type) which, although more resistant to the bacteriophage, is more susceptible to the predatory action of the phagocytes; or, in some cases to transform them into ultramicroscopic and filtrable bodies. If these resistant forms exist in parts of the body in which they can be reached by the phagocytes, they are likely to be destroyed. If, on the other hand, they are not so located, they may establish themselves as the agents of long-enduring, chronic infections, such as those of the gallbladder, urinary bladder, urethra or other organs. These conditions of infection naturally merge with those found in the carrier state, from which the R forms of culture can very commonly be isolated.

The Choice of a Bacteriophage for Therapeutic or Prophylactic Use. —In the actual employment of the bacteriophage in the treatment of specific infectious diseases it has been the usual practice merely to seek a bacteriophage homologous for the normal sensitive culture of the infective agent, and considerable success has attended this practice. But it now becomes apparent the matter may not always be so simple, and it is possible that certain reported failures in bacteriophage therapy may be due to a circumstance heretofore accorded little attention either by bacteriologists or by physicians. Whether one employs a small area or a large area principle may determine important differences in the results attained. I shall first examine the grounds for this view.

In treating specific bacterial infections we are not confronted with the necessity of employing a single form of the bacteriophage in the ordinary sense, but have the choice of two forms. These are the strains which produce, respectively, the large and the small lytic areas. We have seen that the bacteriophage cannot be regarded as a homogeneous unit, for it has these two distinct forms of manifestation; and we may well conclude that the corpuscular elements producing these areas are distinct entities, each produced by a definite form of culture.* For the sake of convenience in reference, I propose that these two forms of the bacteriophage corpuscle should be termed the alpha and the beta units, giving, respectively, the large (a) and the small (β) plaques. As has been seen, there are no intermediates. Some lytic areas which seem to be intermediate in size, as first described by Bail,[26] in reality belong to the a type, in which there is greater relative variation in size than among the β plaques.†

Without again entering into the details of Gratia's interesting experiments with his large and small areas (B. coli), or of the experiments performed by my students and myself on the paratyphoids and other species, the present knowledge of the relation of bacteriophage type to culture type may be briefly and tentatively presented as follows.

The a units, producing the large areas, are regenerated mainly by cultures in the S cyclostage, less readily by the early R type culture, such as that produced by the action of immune serum (after several serial transfers) on the homologous S culture. They may be regenerated readily by S type culture of other strains of the same bacterial species, but not necessarily by the S cyclostage of different, but at the same time related, species. For example, the a units of the antiparatyphoid B bacteriophage do not appear to regenerate at the expense of type S typhoid culture. Moreover, they are not regenerated at the expense of

---

* One might contend that the large and small areas are caused by differences in the cultural substratum reacting to the same bacteriophage; but that this is not true is shown by many tests which I have performed with my students, Eugenia Dabney and B. Jiménez. The details of this work will be presented later.

† The a and the β units are to be found in all bacteriophage filtrates at some moment, provided these have been built up from cultures in the sensitive S cyclostage. If the cultural substratum has transformed in any degree toward the R type the development of the β units is favored; and, if the bacteriophage is continued in serial transmission on such a substratum, the a units may be completely lost. They should be looked for in the filtrates of the first two or three serial filtrations. Then, if one desires to perpetuate the a principle, this must be accomplished through a lytic colony selection, thus yielding a "pure-line" a principle. It cannot be doubted that much investigation on the bacteriophage has been conducted, not with the "complete" lytic agent, but with only the β fraction which manages to survive after the a units have disappeared. It has previously been pointed out that, while the β units are stable and perpetuate in this form, the a units often seem to dissociate into the a and β units. Thus, any a strain must be frequently "purified" by colony selection.

cultures which have arisen as secondaries to the a principle itself. The a units may, however, regenerate from certain secondaries left after the action of the β principle, just as the β principle may regenerate from certain secondaries left by the a principle. This reciprocity of action existing between these two principles, first pointed out by Gratia for B. coli, I have been able to confirm fully for B. paratyphosus B. and its chief dissociates.

The β units, producing the small areas, are regenerated to some extent on normal sensitive culture, but especially on culture of the R cyclostage, such as the resistants produced by the action of immune serum on the homologous S culture. They are regenerated equally on certain cultures which have arisen as secondaries to the action of the a principle on S type culture. Indeed, it appears that the two forms of resistant culture last mentioned are identical, whether produced by immune serum or by the a principle. The β units are not, however, regenerated at the expense of certain secondary cultures which have arisen from the action of the β principle itself. Differing from the a units, the β units may be regenerated at the expense of cultures that are heterologous but closely related. For example, the β units of B. paratyphosus B are regenerated on typhoid culture. It was noted previously that the a units were not so regenerated. If a lytic filtrate containing both a and β units of antiparatyphoid B bacteriophage are brought into contact with a sensitive B. typhosus substratum, apparently only the β units regenerate and produce areas. These are consequently small, although they may be larger than the analogous β areas on paratyphoid itself.

It seems to me that these observations (which, up to the present, we have established only for the paratyphoid bacteria) have an important bearing on one of the significant aspects of serologic convergence and the cosmopolitanism of the R types. I may put the matter in the following form: In relation to convergence it is now known that:

For most bacterial species, at least, there exist two chief types of culture (cyclostages), the S and the R.

The S type culture is characterized by the possession (in the main) of the heat labile, specific antigen.

The R type culture is characterized by the possession of a heat stable, non-specific antigen.

The heterologous, serologic affinities between different bacterial species, when they occur, are maintained by the heat stable antigens, including the R.

Under similar conditions it is also known that:

For many bacterial species (and probably for all) there exist two elementary types of bacteriophage, the α and the β units.

The α units are generated (chiefly) by the heat labile, S type culture, and are themselves heat labile (62 to 63 C.).

The β units are generated by the heat-stable R type culture and are themselves heat stable (72 to 74 C.).

There is some evidence (in B. paratyphosus) that the heterologous, bacteriophagic affinities manifested by a given bacteriophage suspension are determined by the heat stable β units, rather than by the heat labile α units.

There thus exists the inherent suggestion that not only the heterologous affinities of a serologic nature, but also heterologous affinities of a bacteriophagic nature, are dependent on the behavior of either the heat stable bacterial R antigen or the heat stable, β-bacteriophagic units that are generated at the expense of this antigen.

It is possible that these circumstances which agree so well in the main, are only a coincidence; or it may be that they will not be found capable of demonstration in other bacterial species. On the other hand, it is possible that they should be interpreted as carrying the suggestion (which does not lack support from other fields of observation *) that the protoplasm of the bacteriophage (in each case) is, in some way, continuous with the protoplasm of the bacterial cell in which it arose; that certain fundamental characters which dominate the one also reveal themselves in the cardinal attributes of the other.

Summary.—From these observations, the supporting evidence for which will be presented in a subsequent publication, we are led to the conclusion that the "bacteriophage," active on a given bacterial species, is not a homogeneous unit, as usually pictured, but has a dual nature, as sufficiently demonstrated by the diverse manner of action of the α and β lytic units. We observe, moreover, that each of these units is related (in point of regenerative ability) to one or more definite cyclostages, involving not only those of the species of the substratum, but those of related bacterial species. We see, moreover, that, at least within certain limits, the resistant cultures arising from the action of the α principle are influenced by the β principle, and that the reciprocal circumstance also holds true. We come now to a consideration of the possible significance of these facts, gained as a result of test tube experiments, for some of the problems of bacteriophage therapy.

* Note, for example, the relation between Koser's "thermophilic" bacteriophage and his thermophilic culture.

In my earlier work on microbic dissociation it was pointed out that virulence (and often the toxigenic functions) of micro-organisms is often, and perhaps always, the property of a certain stage in the cyclogeny of the species; and that, in many pathogenic species at least, virulence seems to attach mainly to the S cyclostage. We know, moreover, that, in the course of several acute infections determined by S type cultures (leaving aside the possibility that, in some cases, these may be supported by a filtrable stage of the infecting organism), the infecting culture becomes transformed, through the dissociative reaction, into, or toward, the R form which commonly lacks virulence and often shows a much reduced toxigenic power. This point is demonstrated by certain cases of pyelitis or other urinary infections due to B. coli; also by some chronic typhoid infections and by the carrier state. Under these circumstances it might be predicted from laboratory tests that the form of lytic agent employed in therapeutics would be of some importance. Indeed, it seems probable that if any bacteriophage treatment were to prove effective it would involve the use of the β principle. Acute pyelitis, on the other hand, like acute typhoid, should be benefited by the administration not of the β units, but of the a, or of a mixture of the two. In other words, it is suggested that a bacteriophage should be employed that is not only homologous with the species of the infecting microbe, but also adapted to the particular cyclostage that predominates in the infective process.*

That these considerations actually have significance in practice, although they have previously not had adequate explanation, is suggested by the results of several investigators and particularly by experiments of Zdansky. This investigator observed that cultures of B coli (from pyelitis) that were resistant to one strain of bacteriophage might be susceptible to another strain. He did not, however, differentiate these strains other than to perceive their different effects on the infecting cultures. This situation, existing in the natural course of pyelitis and related infections, manifestly finds its explanation in the facts outlined previously, namely, that the organism causing the initial infection (acute or semi-acute) is almost invariably the more virulent S type (Larkum). As a result of normal dissociation, or perhaps under the influence of the a principle, if applied, this culture is transformed, in the body, to

---

* I have been advised by Dr. Cowie, who has had unusual success in the administration of coli bacteriophage in cases of urinary diseases in the University Hospital, that he has often made this observation: A strain of lytic principle which gives only inferior results in test tube lysis may prove more valuable in actual therapeutic work than a bacteriophage strain that gives strong test tube lysis.

an R form, possessing diminished virulence; and this form remains in the tissues producing the chronic infections so typical of pyelitis. In such cases the resistant form is often discovered existing in the presence of the a lytic principle which is unable to influence it appreciably, but which can effect the lysis of normal, sensitive coli cultures. As Zdansky was able to demonstrate, however, a lytic principle capable of further dissociating this resistant culture may usually be produced. Such a bacteriophage is manifestly a principle of the β type.

From these considerations it might seem feasible, for therapeutic purposes, to produce and maintain in the laboratory both the a and the β principles, separately and in high concentrations, employing them separately or mixed, as dictated by the results of bacteriologic tests made from the infected sites. Although much remains to be learned regarding the behavior of the a and β principles and their actual significance in the bacteriophage reactions, I believe it may be anticipated that continued study of them will furnish an explanation of some of those instances in which the administration of the lytic agent has yielded negative results, and will thereby make possible further improvements in the methods of bacteriophage therapy.

The Bacteriophage as a Test Agent in the Study of Respiratory Transmission in Filtrable Virus Infections.—Although d'Herelle's hypothesis regarding the existence of an "infectious immunity" based on the dissemination of the bacteriophage from person to person, during the latter stages of an epidemic, has not received any substantial support in recent years, there is an aspect of transmission of the agent from person to person, under experimental conditions, which possesses considerable interest. This point involves the practical use of the bacteriophage in studying some of the conditions that might surround the droplet transmission of a possible infection by a filtrable virus. Whether the lytic agent is or is not a filtrable virus, it must be regarded at present as a living agent lying, for practical purposes, as close to a filtrable virus as is likely to be available for such a study for some years to come. It possesses the advantage of making possible a laboratory study of the conditions surrounding the transmission of a minute, living, filtrable and ultramicroscopic particle. Such a study was made by Olsen and Strauss [265] in Berlin in 1926.

These investigators, employing a Shiga bacteriophage for contaminating the mouth of one of their subjects, studied the distance to which the agent might be propelled in droplets expelled from the mouth during talking or coughing. Of greater interest, however, were their experiments involving the direct transmission of the bacteriophage corpuscles from one person to another by the same means. The experiment was as follows: Subjects A and B stood opposite and facing each other 50 cm. apart. The mouth of A was infected with bacteriophage filtrate. Subject A then spoke somewhat loudly for two minutes into the face of subject B. The latter, during this period of exposure, breathed alternately through nose and mouth in ordinary respiration. At the conclusion of the test period, examination of the nose and mouth of subject B revealed abundant bacteriophage. The nasal secretions and saliva were diluted with sterile broth. They were then spread on agar plates which had previously been coated with the sensitive culture. After appropriate incubation, numerous lytic areas appeared. From these experiments it appears that an ultramicroscopic and filtrable agent may readily be transmitted by droplet infection from one person to another, merely through the process of energetic conversation. In this manner, the bacteriophage lends itself, perhaps better than B. prodigiosus, to many interesting aspects of experimentation in which it may be employed in the place of an actual ultramicroscopic virus.

Conclusions.—In concluding this phase of the subject it may be said that, although probably not meeting all of d'Herelle's early expectations relating to a far-reaching, and even communicable, endogenous immunity attributable to the bacteriophage, the evidence gained within recent years fully supports his conclusion regarding the therapeutic (and, perhaps in a different way, the prophylactic) value of bacteriophage administration in several important infectious diseases, as well as in suppurative processes. It seems to me, however, that the significance of the immune reactions now observed, while on the one hand scarcely meeting d'Herelle's expectations, have on another side revealed to us a mechanism of defense different from anything d'Herelle has conceived; and vastly greater in its range of manifestation and possible usefulness. Here I refer once more to my conception of bacteriophage reactions as a single aspect of the phenomenon of microbic dissociation; and of the latter, when it occurs in vivo, as a phenomenon of extreme significance in the defense mechanism of all animals.

Omitting, for the moment, the relation of the bacteriophage to the immune reactions, I may briefly recall some of the points which I have developed in a former paper—to the effect that the immune reaction, so far as it concerns the destruction of virulent organisms in the body, is essentially concerned with microbic dissociation in vivo. Virulent and nonphagocytable bacteria are transformed into nonvirulent and phagocytable forms. This view was first introduced by Griffith on the strength of his splendid study of pneumococcus dissociation and immunity published in 1923. When one stops to consider the mechanism of this transformation, ample evidence appears from experiments in vitro that it concerns the action of immune bodies probably identical with the bacteriotropic antibodies or opsonins. Therefore, it seems possible to look on the chief function (the so-called "preparing action") of these opsonins as that of enforcing, in vivo, the dissociative reaction.

But I have also attempted to point out in the course of this review that the fundamental significance of the bacteriophage reactions involves, not lysis and final destruction of the sensitive bacteria, but a dissociation in which the original culture of the substratum is forced into a series of transformations terminating in the generation of the secondary, resistant types—themselves nonvirulent and susceptible to phagocytosis. Thus, as it seems to me, the instigating corpuscles of the bacteriophagic dissociative reaction come to possess a protective function. If numerous and "virulent," they may effect an immediate protective action by "lysis" of the invading bacteria; if less numerous, or less "virulent," their action can be detected in the slower dissociation-furthering process, terminating in the production of nonvirulent and phagocytable cell types. In other words, it seems that the bacteriophage and the specific immune serum accomplish exactly the same end, so far as the nature of their influence on the virulent culture is concerned; only the former accomplishes its end, sometimes with startling rapidity, while the latter acts more slowly. In both reactions, however, the dissociation of the invading organisms results. It was for this reason that, when I earlier [168] recounted some of the "incitants" to microbic dissociation, I said that there were two which stood out above all others in the speed of their action and the effectiveness of their results—namely, the bacteriophage and homologous immune serum. What the former accomplishes directly, the latter enforces the bacteria to do for themselves.* It will be seen later that there is evidence that the mechanism is the same in both cases.

Finally, attention may be called once more to the circumstance that the therapeutic administration of the bacteriophage involves a double reaction. While the action of the lytic agent itself is undoubtedly direct, that arising from the dissolved bacterial substance or bacterial residues is mainly of opsonic significance. Whether a "pure" antilytic serum

* It is perhaps not without significance that Hauduroy [174] observed the presence of typhoid lytic agent in the blood of the majority of his typhoid patients at a certain stage of the malady, and that Sonnenschein [815] made a similar observation in patients with paratyphoid.

might possess opsonic power is still unknown, but present evidence (Arnold and Weiss,[16] Weiss [331]) would suggest that it does not. Thus, the end-result of bacteriophage action, per se, and the end-result of administering bacterial products capable of generating a strong, opsonizing serum are the same: Both serve to evoke the dissociative reaction, through which the invading organisms, if of the sensitive type, are transformed into harmless, or sometimes only less harmful, forms.

Which of these two aspects of the dissociative mechanism is of greater importance in a given case, only further study will reveal. The reports thus far available suggest the greater significance of the direct action of the bacteriophage in therapeutic endeavors, and of the indirect action of the lytic filtrate as a whole in prophylaxis. So far as is known at present, however, there is not any disadvantage in making use of the combined agencies in a single filtrate. In future work, either through securing a more perfect "adaptation" of the principle to its substratum, the infecting culture, or through the use of mixed bacteriophagic suspensions of a "polyvalent" nature, it may be anticipated that further improvement will be effected in lytic principle therapy. As the new work progresses, moreover, it seems probable that the field of lytic principle therapy will tend to merge more and more into the field of Besredka's antivirus, the exact mode of action of which is still unrecognized. That the two possess a common factor is already strongly suggested.

9. VIEWS ON THE NATURE OF THE BACTERIOPHAGE AND OF THE MECHANISM OF LYTIC ACTION

In the entire field of investigation on the bacteriophage probably no subject has awakened more discussion than the nature of the agent itself. Much of the study has involved the question of its living or nonliving nature and, incidentally, has served to reveal the great difficulties inherent in attempts to interpret evidence bearing on this subject. The problem is certainly complex, and is rendered even more inapproachable by the circumstance that certain observed facts are in harmony with views of highly diverse nature.

There are few writers on the subject of the bacteriophage who have not expressed an "opinion" regarding its nature. But such opinions have usually been based on evidence dealing with only one phase of the problem, or concerned with only one or two of the attributes of the principle. Other investigators have made a closer study of the diverse factors of the problem. The opinions of such workers are entitled to special consideration. It may be added, however, that the final conclusions regarding the nature of the bacteriophage will not be derived from the results of any single test, or by the application of any single criterion of living stuff—much less by the use of applied logic or of formal definitions as to the attributes of living protoplasm. Final conclusions can be drawn only after a thorough analysis of many diverse factors; and for a long time to come may be little better than the establishment of a probability.

REFERENCES

The Bacteriophage as a Living Agent

A. Without definite statement as to the nature.—Allesandrini;[1] Hauduroy;[182] Kuhn[221a] (either a stage in the development of the bacteria or a virus (Myxomycete) living in symbiosis with the cells).

B. Foreign to bacterial cell (see also d'Herelle[188 4]).—D'Herelle[188] (filtrable virus parasitizing bacteria); Bruynoghe[67] (same); Asheshov[27] (same); Schuurman[301] (same); Preisz[266] (same); Marshall[246] (same); P. C. Flu[155] (same); Kuhn[222] ("spores" of a myxomycetean parasite of bacteria); Koch[215a] (same as Kuhn).

C. Biologically related to bacterial cell.—Bail[30 35] ("splitter theory"; part of generative substance of cell); Hoder[197] (same); Enderlein[117] (symplastic stage of bacterial cyclogeny); Wollman[355] (parahereditary transmission; function of cell reproduction); Wollman and Wollman[347, 348] (same); Rosenthal[297] (related to bacterial sporogony); Bechold[254] (bacterial fragments possessing enzymatic power); Otto and Munter[304] (similar view); Wollmann[356] (hypothesis of "'facteurs' héréditaires"); Hadley[168] (homogamic theory; bacteriophage action a phase of microbic dissociation).

The Bacteriophage as a Nonliving Agent

A. A chemical principle foreign to cell.—Kabeshima[205, 206] (prodiastase exciting bacteriolytic function; Kuttner[220] (enzyme from tissue cells).

B. A chemical principle derived from bacterial cells.—Bordet and Ciuca[42] ("nutritive vitiation"); Bordet[40] (same); Lisbonne and Carrère[234] (bacterial antagonism); Seiffert[310] ("exogenous autolysis"—catalyst theory); Doerr[108] (toxin regenerated by diseased bacteria); Davison[104] (normal autolysis); Carrell[85] (a substance liberated from the cells); Kasarnowsky and Timokin-Schuckoff[208] (lysins composed of bacterial substance); da Costa Cruz[94] (a globulin from the cells); Prausnitz[266] (ferment nature suggested); Pico[261] (the activating factor of a normal autolytic principle); Otto and Munter[304] (a lysin arising from the disintegration of the bacteria); Gildemeister[137, 138] (a normal autolysin); Zinsser[347] (an enzyme); Kuttner[225] (an autolysin liberated from bacterial cells and acting as a catalyst); Arnold and Weiss[24 a] (a normal autolysin); Eastwood[112] (disturbance of synthetic-catalytic balance); Okamoto[201] (a nitrogen-splitting ferment); Fejgin[124, 129] (a normal autolysin); Borchardt[34] (same); Bechold[254] (minute bacterial fragments arising from the disintegration of bacteria and possessing enzymatic power).

Other Views on Nature or Origin

Otto and Sukennikowa[271] (related to paragglutination).

Breinl and Hoder[51] (relation to bacterial mutation).

Gohs [140, 142] (dystrophy theory).

Schnabel [300] (anaphylactic theory).

Kauffman [210, 211] (relation to microbe dissociation—"Keimunwandlung").

Hoder [197] (relation to mutation phenomena; favoring Bail's theory).

Eastwood [112] (relation to bacterial variation).

Béguet [31] (relation to variations in osmotic pressure within the bacteria).

Proca [280] (amebic theory).

The Heterogeneous Nature of the Lytic Units

Bail [20] (an "elementary species" of the bacteriophage).

Bail and Watanabe [27] (on "elementary species of the bacteriophage").

Watanabe [328] (serological aspects of the "elementary species").

Gratia [160] (heterogeneous nature of a coli bacteriophage).

Gratia and de Kruif [161] (heterogeneous nature of a coli bacteriophage).

Hadley [169] (large and small area principles in a Shiga bacteriophage).

Kline (heterogeneous nature of typhoid bacteriophage).

Hadley and Dabney [169 a] (dual nature of an anti-paratyphoid bacteriophage; the alpha and the beta units).

DISCUSSION

The conception of d'Herelle regarding the nature of the lytic principle is too well known to demand extended treatment here. Briefly, it concerns a foreign, filtrable virus (Protobios bacteriophagus, synonym, Bacteriophagum intestinale d'Herelle, 1918), which parasitizes bacteria and causes their destruction by lysis. Bacteriophagy is, therefore, a "contagious disease of bacteria." The infectious unit is an ultramicroscopic and filtrable corpuscle, multiplying only at the expense of young, living bacteria, some of which may acquire an "immunity" to the parasite, and thus become resistant to the lytic action.

Theories Proposed.—The first investigator to express doubts of d'Herelle's view was Kabeshima. Basing his conclusions on the resistance of the principle to chloroform and other antiseptics, this worker reached the conclusion that lysis was due to the action of a sort of catalyzer produced by the digestive glands (perhaps by the leukocytes) which activated a diastase normally present in the bacterial cells. This view has had few adherents. D'Herelle has already called attention to its insufficiencies. Moreover, one now can see that the influence with which Kabeshima dealt was, in reality, only the so-called "liberating influence" which has more recently been carefully studied by Klimek and myself, as well as by other investigators. I have repeatedly pointed out that there must be a clear differentiation between the principle which liberates the lytic reaction and the principle which underlies the serial transmission of the phenomenon; one is a condition or influence, while the other is a physical unit.

In 1920, Bordet and Ciuca interpreted the bacteriophagic reaction as an autolysis or bacteriophagy, resulting from the rupture of the equilibrium existing between assimilation and metabolism (nutritive vitiation). According to their view, this vitiation has its origin in a normal physiologic characteristic of bacteria; but, under certain conditions, or at the instigation of certain bacterial cells, it may acquire a pathologic significance in the life of the cell or culture. After certain bacteria have experienced this vitiation, they may communicate the condition to surrounding sensitive cells by the products of their autolysis; and the reaction thus may be transmitted in series. Bordet thus considers the possibility that the bacteriophage may not itself possess a lytic function, but that it is the excitor of autolysis.

Bordet regards the microbes in any culture susceptible to lysis as being variously receptive to the exciting substance—a point of great significance which has not been sufficiently emphasized by d'Herelle. As a result of this, a selection occurs in which some bacteria qualify as resistant forms and therefore do not reproduce the lytic agent, at least actively. Others succumb to lysis while at the same time they regenerate the principle; still others may regenerate it without appearing to experience any deleterious effects, often giving luxuriant cultures. In this case the resistant forms manifest an equilibrium between the aptitude to create new, living substance and the tendency to auto-destruction. "They are in some way in harmony with the principle which they elaborate, and against which their associates (which have not attained this stabilization) show themselves still susceptible."

The cultures which have made the adaptation, Bordet points out, are also modified in other ways; and thus it is that the lytic principle functions in the production of certain new microbic types and assures the maintenance of such forms. "The essential character of being reproducible, which it possesses, guarantees the permanence of its action through successive microbial generations." Thus, according to Bordet, it is the lytic agent that safeguards the stability of the characters that define the living species or variety, in the course of repeated cell divisions. The chemical principle which controls these characters is incessantly produced, and always in quantity sufficient to prevent the descendants from departing from the specific type. Thus, for Bordet, the bacteriophage reaction becomes an exaggeration of a transmissible autolytic tendency, manifested to a moderate degree by quite "normal" bacteria. The transmissible aspects can appear clearly only if the principle studied possesses unusual activity, or acts on unusually sensitive cells.

But it is equally important to note how, according to Bordet, the principle originates. In 8 of 12 attempts, he obtained it through the action of the leukocytic exudates of guinea-pigs after intraperitoneal inoculation of cultures of B. coli. Therefore, his view that the exudate "releases the lytic action"—a notion entirely in harmony with much later work, although not necessarily limited to experiments dealing with the influence of leukocytes alone. The important question is, What is "released"? Assuming that the autolysis so generated in the cultures was not due to a living parasite, and believing that it is not within the power of experimentation to confer on a living organism a really new property, Bordet believed that the autolytic tendency observed must represent the exaltation of a "normal property." In the peritoneal cavity of the guinea-pig, surrounded by leukocytes, the bacteria undergo a selection which results in the survival, and later proliferation, of certain microbes capable of acquiring the "special lysogenic aptitude." The products of these cells are then able to release the same process in normal cultures when the guinea-pig exudate is added. The irregularity with which the phenomenon has been found to occur Bordet has explained as due to the fact that the "special" microbes are rare, or sometimes entirely lacking, in cultures.

It will be seen later that there is reason to suppose that, but for one or two departures (attributable to certain remnants of the monomorphic point of view which dominated his conceptions of the nature of the variable forms that he clearly observed playing a rôle in the lytic reaction), Bordet approached close to several of the most essential phenomena of the reaction, much closer, indeed, than d'Herelle with his virus theory. With reference to these two opposing views, their respective failures, according to my conception, may be summarized by the simple statement that d'Herelle failed to see or to understand the importance of cell and culture variation, while Bordet, although to a limited extent he noted its importance, gave to it an erroneous interpretation.

The three theories of bacteriophage action just mentioned have furnished the chief basis for all the more important discussions up to the present. While d'Herelle's theory is too concrete and specific to receive significant modifications at the hands of other investigators, and while the views of Kabeshima have gradually lost the support which they at first received, the theory of Bordet and Ciuca, either in the original form or with some modification, has been accepted by the majority of workers. Accordingly, the bacteriophage has commonly come to be regarded as an autolysin, arising in the bacteria themselves. Other conceptions, such as the "splitter theory" of Bail, the anaphylactic theory of Schnabel, the dystrophy theory of Gohs, the sporogony theory of Rosenthal, the Pettenkofer cell theory of Kuhn, the amebic theory of Proca and the symplastic theory of Enderlein, have had few or no adherents. The same is true of my homogamic theory, briefly presented in 1927, although several investigators (Kuhn, Breinl and Hoder, Hoder, the Wollmans, Eastwood, Kauffman) have shown distinct tendencies to approach, from one avenue or another, the fundamental biologic conception of the relation of the bacteriophage phenomenon to cell variation and "mutation." I much regret that the interesting theory of Kuhn was not referred to in my earlier review. This will be considered subsequently.

In view of the prominence of the theories of d'Herelle and Bordet, and the wealth of experimental support which each has given to his own hypothesis, I will now consider these in the light of more recent investigation. The problem here resolves itself into consideration of the bacteriophage (1) as a filtrable virus which parasitizes bacteria, or (2) as an autolysin (chemical substance) reproduced by the bacterial cells. These questions, in turn, resolve themselves into the problem of the bacteriophage as a living, or as a nonliving, factor; also as a substance possessing a physical form (corpuscular), or a substance represented more accurately by a chemical product in the state of solution.

Physical Nature of the Bacteriophage.—Before considering its nature from the point of view of a living or a nonliving factor, I will take up the question of its probable physical nature. Here, from the beginning, there have been two opposing views. The first is that of d'Herelle who maintains its corpuscular nature in the sense that it is a distinct physical unit, although ultramicroscopic and filtrable. The second is that of Bordet and Ciuca who have maintained that the agent is something in solution which can be measured only by its "strength," and in which the agent does not exist in structural units or corpuscles. These views may now be considered in greater detail.

D'Herelle's conception of corpuscular units is supported largely by the possibility of enumeration of the lytic principle, and by certain correlated phenomena. When a suitable dilution of a lytic filtrate is streaked on agar which was previously coated with some of the sensitive bacteria, the agent is found to register by the production of lytic areas or plaques appearing as bare spots against the background of culture. If a larger amount of the same lytic filtrate, or the same amount of a lower dilution of the filtrate, is employed for the tests, the number of plaques, other things equal, will be correspondingly greater. Similarly, the number of plaques may be reduced by using a smaller amount of the same dilution of the filtrate, or the same amount of a higher dilution. An exact correspondence between the amount or the dilution of the filtrate and the number of areas is not always observed; but, certain factors of error being taken into consideration, the correspondence may be regarded as close, as has been shown by many investigators.

For d'Herelle, each of these plaques represents the site of action of a single lytic unit which has multiplied at the expense of the surrounding culture, and thus forms a "colony" of lytic corpuscles. The number of the lytic colonies thus becomes an index of the number of lytic units present in the original suspension.

As I have already briefly observed in another connection, another explanation of the plaques is found in the views of Bordet and Ciuca. According to them, the number of areas is dependent, first on the "strength" of the filtrate; secondly, on the presence, in the culture of the substratum, of a certain number of cells possessing such a degree of "sensitiveness" that they can be influenced by the concentration of the active substance in the filtrate. As the strength of the filtrate is increased, correspondingly larger numbers of cells possessing the necessary degree of sensitiveness are found; and, in a similar manner, as the strength of the filtrate is decreased a correspondingly smaller number of cells sufficiently sensitive to be influenced is found. Thus, for Bordet, the strength of a bacteriophagic filtrate is measured, not by the number of lytic corpuscles contained in it, but by the number of sensitive cells in the substratum, on which it can impress itself.

In such a difference in interpretation of the essential facts relating to the appearance of the lytic areas, one might anticipate that the question is open to experiments involving the use of different numbers of cells and identical amounts of lytic filtrate; or the same number of cells and different amounts of filtrate. In such tests as those performed by d'Herelle it seems to have been clearly shown that the number of plaques is not modified appreciably by the number of cells submitted to the action of the agent concerned, as should be the case if Bordet's view were correct. The experiments of d'Herelle have been confirmed by Collins [90] and by Marshall [246] in this laboratory; also by Asheshov and many other workers, and most recently and clearly by Wollman.[336] I do not see that Bordet's view of this matter can actually be disproved

—at least in any way other than by showing that the evidence supporting another interpretation is far stronger. And this, I believe, is the
present situation with respect to the lytic areas. Such evidence as is
now available certainly supports the conclusion that the bacteriophage
possesses a definite corpuscular form; that it is a physical unit in the
same sense that this characteristic is attributed to the ultramicroscopic
and filtrable stages of known bacteria. It seems to me that the interpretation of Bordet is in harmony with the older views of the nature of
bacterial autolysis—the vanishing of the living protoplasm into soluble
stuff. But, as I have tried to point out in an earlier publication, the
so-called "lysis" of bacteria may conceal more that one type of reaction;
and I shall have occasion to deal with this subject again later in the
present work. (Section 12).

Although the assumption that the bacteriophage possesses a physical
form naturally carries with it the implication that this is a living form,
it is desirable to examine the additional evidence supporting this view.

Living Nature of Bacteriophage.—For d'Herelle the living nature
of the bacteriophage is demonstrated by its power of adaptation and
power of assimilation in a heterogeneous medium. If, as he believes,
these characteristics are attributes of living things only, then, by force
of logic, the bacteriophage is living, for it certainly seems to assimilate
and propagate itself in a heterogeneous medium, namely, the bacterial
cell. But there may be involved here a fallacy, in the mode of argument
if not in the conclusion. I believe it is questionable whether any one
has yet actually demonstrated that the bacteriophage "assimilates" and
"propagates itself" in a heterogeneous medium; and it is not necessary
that it should do so in order to possess attributes of a living thing.
Is the sperm cell of the mammal, the microgamete of the protozoan, the
pollen grain of the higher plant or the fertilizing cell of the fungus a
living cell? Or do they "propagate" in a "heterogeneous medium"?
They are not autonomous living cells, but all function physiologically;
and they serve eventually to generate, as a rule, other similar bodies,
although there is an alternation of some cell generations between, just
as one knows that some cell generations exist between the "attack" of
the bacteriophage and the liberation of the new brood of corpuscles. The
criteria which d'Herelle seeks to apply lose some of their significance,
moreover, when one considers the circumstance that just what happens
in the bacterial cell during that interval lying between the "invasion"
by the lytic corpuscle and the generation or liberation of the new brood
of corpuscles is not known. There is no evidence, as yet, to assure one

that the generation of the lytic units is a continuous and uninterrupted process, carried on by simple fission within the host cell, as the theory of d'Herelle demands and implies.   There is, on the other hand, some evidence to show that the first response to the "invasion" of a culture by the bacteriophage is unusually rapid growth, recognized by d'Herelle and by Bordet, and termed by the latter "proliferation," because of its luxuriance.   This can mean only that the bacteriophage, or its influence, must be carried along from cell to cell with each new division.   Here it would then appear that the misfortunes of the parents are visited on the children and grandchildren to the nth generation.   This might well be termed a cumulative mode of parasitism.   I believe, however, that there is a better explanation, which will be mentioned again.

The apparent "power of adaptation" of the lytic corpuscles to work successfully under conditions at first adverse might be regarded as additional evidence for their possession of a distinct biologic significance. And this has often been alleged.   It must be borne in mind, however, that the lytic agent cannot be revealed in action in an isolated state; it must always be accompanied by the culture of the substratum, through whose behavior, not only the presence, but also the characteristics, of the bacteriophage are revealed.   From many experiments involving this joint action it has been concluded that the bacteriophage accomplishes the adaptation.   I do not believe that these conclusions are justified, although they have never been disputed.   The nature of the data reported in such experiments almost invariably demonstrates just as clearly that it is the bacterial culture which adapts; and that the new ability is transmitted to the fresh culture through the medium of the bacteriophage.   This point, which concerns a new theory of bacteriophage action that will be presented in a later section, will probably not be clear at this stage.   Suffice it to say for the present that the assumed adaptation of the bacteriophage, as a corpuscle biologically foreign to the bacterial cells, is interpretable on other grounds than those most commonly brought forward.

Regarding the evidence of the marked resistance of the bacteriophage to chemical influences and to germicides—taken by many investigators as an indication that it cannot be a living thing—here the results are difficult to interpret on the basis of any present knowledge.   Although living organisms which are able to withstand many of the agents that have been shown to have no or slight influence on the bacteriophage are not known, the knowledge of the resistance of endospores and arthrospores should cause one to hesitate in concluding that the bacteriophage cannot be a living thing merely because of its superior resistance to lethal agents. And this may include heat, light, oxygen, enzymes, radiation, immune serum and probably other factors. None of the experiments already reported dealing with these agents militates against the view of the living nature of the bacteriophage. Indeed, its reactions against some of them (particularly heat, enzymes and radiation) are such as to favor this view. As will be seen later, however, the conclusion that the lytic agent is living does not compel acceptance of the view that it must be a filtrable virus in the d'Herelle sense. There is another possibility—one lying as far outside Bail's "splitter theory" as it does outside of d'Herelle's theory of the ultravirus.

Kuhn and the Pettenkofer Bodies.—Before approaching this new possibility, however, I wish to present certain observations and conclusions of an investigator whose work, in the annals of bacteriophage study, seems earlier to have remained unknown. It has never been referred to by Bordet nor by d'Herelle, so far as I am aware, and was not included in d'Herelle's 1926 bibliography.* The particular article that I refer to is one by Kuhn [221] of Dresden, and carries a title that might easily conceal its true significance—"Weitere Einblicke in die Entwicklung der A-Formen (Pettenkoferiaformen)." In the form presented by the author it can scarcely be said that the subject matter of this paper justified the phrase, "a theory of the bacteriophage," and it seems clear that Kuhn did not present it as such. But the nature of the facts revealed, the manner of interpretation and the brief hypothesis formulated, are sufficient, I believe, to justify a consideration of this work as a theory of the bacteriophage; and as a theory which, to my mind, approaches the mystery of the bacteriophage more closely than has yet been possible from any other avenue. And this is particularly true when one interprets Kuhn's results in the light of the dissociation reaction, as can now readily be done. I have no hesitancy, therefore, in including in this section, along with the theories of d'Herelle, of Bordet and Ciuca and of others, the theory of Kuhn, based on the reproductive behavior of the Pettenkoferia (the A forms), and worthy of the closest attention on the part of bacteriologists interested in the bacteriophage phenomenon.

The observations of Kuhn presented in 1924 and 1926 had their roots in earlier studies published in 1919 and 1922 dealing with the various types of cells present in "normal" cultures, and their correlation

* Neither was it included in my 1927 bibliography on microbic dissociation, for it is only recently that my attention was directed to these valuable papers (references 221, 221a, 221b

with certain culture types. Even in 1919 Kuhn [221a] made the prediction that the bacteriophage was either a stage in the development of bacteria or a virus living in symbiosis with the bacterial cell. Kuhn (1924) postulated four cell-forms which he designated A, B, C and D. The D type (dendritic) was predominantly a thread form, the filaments often having a large diameter as a result of lateral fusions. They stained dark with Giemsa and manifested a false branching. The C type (coccus) was composed of small round forms and arose from "disintegrating" bacilli or vibrios. It seems possible that many of these were identical with the gonidial forms of Enderlein, Mellon, Almquist and others. Kuhn's B type included the common rod and spiral forms which are regarded as the "normal" type of rod or vibrio and which, as is now known, represent the S form of culture. Kuhn's A forms deal with cell structures possessing extreme interest. These, as Kuhn pointed out, are analogous to what have usually been called "degeneration" or "involution" forms. They have been observed by bacteriologists for many years, but seldom studied; indeed, it has been commonly believed that they were incapable of further propagation. In more recent years, bodies of a similar nature have received study especially at the hands of Almquist, Löhnis, Enderlein and Mellon.[352]

These cell forms seemed to arise either on or in the rods, but underwent rapid development accompanied by the gradual disintegration of the rods in which they arose. Eventually the bodies were freed. Next, according to Kuhn, who followed these changes step by step in Vibrio metchnikovi, there occurred a fusion of some of these cells, usually between one that was larger and lighter and one that was smaller and darker. The cell resulting from this fusion grew to an astonishing size, manifestly resembling those pictured by Mellon and others, and clearly observed in several species by me. It was these forms to which Kuhn gave the name Pettenkofer bodies. Some of them appeared as balloon-like structures which, in time, disintegrated and liberated multitudes of minute refractile bodies which could be stained by the Giemsa method, and of which Kuhn [221] presented clear microphotographs. An apparently similar phenomenon was recorded by Hort [352] for B. typhosus many years ago, and Mellon [351, 352] has described a similar "bursting" for certain cell structures observed among the diphtheroids. I have observed the balloon-like forms repeatedly in the Park 8 strain of B. diphtheriae, undergoing dissociation, and have noted the apparently multinucleated condition of the chromatin. I have also often seen the large, empty "sacs," but I have never been able to observe the actual process of liberation of the contents. Judging from Mellon's description, it occurs suddenly. In the Shiga bacillus my student, Klimek, has easily demonstrated the presence of similar elements arising under the influence of lithium chloride. Chromidial elements were often observed in the large bodies, but the liberation of the "spores" was never seen.

The Pettenkofer bodies were observed by Kuhn to occur particularly in the papillae (secondary colonies) growing in the mother culture of various species. He also stated that these cells might be increased by growing the culture on a medium containing lithium chloride, as first demonstrated by Maassen. A temperature of 22 to 28 C. was found to be better than 37 C. for their development. In broth, the bodies tend to sink to the bottom of the tube, dragging with them other culture elements that may be present. Thus, the medium becomes cleared. If the broth contains lithium chloride, it never becomes turbid, even though liberally inoculated with the Pettenkofer forms.

But the point of special interest in the lithium chloride cultures is as follows: From the washings of cultures grown on lithium chloride agar, Kuhn was able to obtain the lytic principle; but he did not succeed in this when he employed washings of cultures grown on ordinary agar. These filtrates containing the bacteriophage could be obtained easily from the washings of cultures grown for seven hours on the lithium chloride medium. From these results, Kuhn concluded that the formation of Pettenkofer bodies in some way favored the production of the lytic agent. The presence of these bodies prevented the normal evolution of the bacteria. Even in 1919 he voiced the opinion that the lytic principle either was a developmental stage in the life of the bacteria, or was a living thing existing in symbiosis with the bacteria. He suggested that, if the conditions in the culture are right for the development of these bodies, then there arises in these cultures a sort of *societas leonina*. A point of further interest was mentioned by Kuhn,[221] citing Loele, to the effect that certain strains of culture can be split off from the Pettenkofer forms which are "charmed" against the further appearance of the destructive or inhibiting influence of the Pettenkofer bodies. In these one can now easily detect the resistant or R type cultures. Regarding the mode of functioning of the Pettenkoferia, it seemed possible to Kuhn that they contained living forms, perhaps minute spores of some sort, which might "fasten upon" the bacteria and increase at their expense through division. Just what part, if any, these bodies played in bacterial reproduction remained a problem.

In a paper appearing in 1926 Kuhn [222] presented other data regarding the Pettenkofer bodies and their assumed relation to the bacteriophage reaction. Here he described more fully how the minute "spores" (measuring from 0.12 to 0.1μ) were liberated from the Pettenkofer bodies, and seemed to "attack" other bacteria and consume them. He observed the "spores" both in and on the normal rod forms, where they seemed to grow at the expense of the bacteria. In these phenomena he could picture a new sort of micro-organism which is present along with the bacteria (symbiosis) of the culture. While, in his 1924 paper, Kuhn had debated whether the Pettenkofer bodies might represent a stage of bacterial development or an independent living being, he now states that he is not in a position to grant the former possibility. He rather inclines to the view that the organism is a new type of parasite, resembling the Chlamydozoa; and that its systematic position is close to the rhizopod-like organisms. This he concludes from his observation, in stained preparations, of the long pseudopodial processes stretching out from the Pettenkofer bodies.

From this later report it appears that Kuhn, while still presenting data that are in close accord with the dissociation hypothesis, is, in his interpretation and conclusions, turning toward d'Herelle's parasitic view. Fortunately, the value of his observations is not lessened for this reason. The fallacy in his parasitic hypothesis, it may be anticipated, will eventually be shown by the fact that, to accept it, one will need to acknowledge the constant symbiotic existence in all bacterial cultures of the parasitic chlamydozoan; for it can scarcely be doubted that all bacterial species, at some time in their cyclogeny, are capable of revealing the bacteriophage reaction, either in the classic, or in a modified, form.

Most recently there appeared a brief, preliminary account by Koch and Ziegenspeck of observations also dealing with the Pettenkofer bodies.* The great importance of their observations, confirming as they do those of Ph. Kuhn, demands a presentation of the essential facts, although many of the details will have to await the appearance of their complete paper.

In cultures of B. subtilis and some other spore formers, examined by a special dark-field apparatus, there appeared, along with the bacteria, certain minute, delicate, granular bodies, possessing vital movement as opposed to Brownian motion. These bodies were seen to penetrate the bacteria, inside of which they underwent further development. Gradually, the refractility of the invaded cells was lost, the contents seemed to

---

* The complete paper is promised for a forthcoming issue of Botanische Archiv, 1927, 19, 275.

disappear and the rods assumed the appearance of shadow forms. From these, a brood of motile granular bodies was then liberated. Certain other granules in the cells were observed to enlarge to globular form, to lengthen out, to constrict and, eventually, to divide, thus producing in each cell two new globular bodies, quite distinct from spores. These also were later observed to escape from the bacterial membrane and then to unite with the minute granular bodies in a manner suggesting fertilization of the former by the latter. These products of the union gradually enlarged to attain considerable size and showed ameboid movements. Eventually they became surrounded by a firm membrane and gave rise, internally, to numerous refractile granules. After a period of rest the outer membrane disappeared, thus liberating into the medium multitudes of minute granules similar to those described at the beginning. Hence, a cycle was apparently completed. These minute bodies were reported as being able to produce the lytic reaction. Further details will be awaited with interest.

What is the significance of these phenomena? Are the minute granular bodies living things? Is a sexual process present? Do the granular bodies belong to the bacteria or to a foreign parasite? Although further experiments left the issues in some doubt, the authors lean strongly to the view that they represent a real parasitic invasion of the bacterial culture by a foreign agent; further, that this parasite belongs to the myxomycetes, as was first suggested by Kuhn. In general, the observations are regarded as supporting d'Herelle's view of the bacteriophage as a foreign filtrable virus, parasitizing and destroying the bacterial cells.

Although a full discussion of this work may profitably await the appearance of the complete paper, the preliminary account is sufficiently definite to warrant certain criticisms. As has been said in the discussion of Kuhn's theory, some of the observations presented by Koch and Ziegenspeck are entirely in harmony with phenomena observed in the dissociative reactions of bacteria of many species, and it seems to me that they can be more easily explained on that basis. The new points which these authors bring out (origin of the minute, granular bodies inside the bacterial cells, copulation between granules, etc.) are more likely to represent some of the still missing features of the dissociative reaction than to concern the cycle of a new kind of living being living in symbiosis with the bacteria. It will be noted that these results confirm, in a measure, certain observations made by Enderlein and interpreted by him as representing sexual reproduction in bacteria.

If, however, one is to believe that all instances of the bacteriophage reaction are cases of Bacteria-Myxomycete symbiosis, with the parasite gaining the upper hand and causing the destruction of the invaded organisms, certain difficulties are encountered. Some of these may be enumerated.

Since it begins to appear that most any "normal" S type culture, either by itself or under the stimulus of certain incitants (sterile), is susceptible of generating the bacteriophage de novo, this circumstance would need to imply that practically every laboratory culture is "infected" with the parasites. And this may well be doubted.

Again, such a view would need to imply the existence of two types of myxomycete parasite producing the infection: one acting on the S and the R^2 type cultures, the other on the S and the R^1 type cultures. Moreover, that each type of parasite is able to attack and to destroy those bacterial cells which had already proved resistant to the other form of the parasite. Further, that each form of the parasite is characterized by a different thermal death point, possesses a different antigenic constitution and produces plaques of different size through its invasive excursions into the culture substratum.

Finally, one might inquire why it is that a bacterial culture, previously not revealing any evidence of myxomycete infection, can be made to do so quickly and vigorously, merely by the addition of such substances as homologous immune serum, lithium chloride, sterile pancreatin solution, sterile trypsin solution, sterile peptone or sterile distilled water, to the young culture? But perhaps these matters will be explained in the complete work of Koch and Ziegenspeck, yet to appear.

At present, however, these and other difficulties that might be mentioned are serious obstacles to the acceptance of Kuhn's myxomycete theory, which is now apparently gaining some ground in Germany. This notion, however, seems, in final analysis, to be merely one more ineffectual attempt to escape from the logical and now fully apparent explanation of bacterial pleomorphism; to explain "involution forms," not as "normal" stages in the cyclogeny of the species, nor yet as "degenerated" cells (as has long been the common practice), but as cells under the spell of a myxomycete infection. I believe that the pure observations of Kuhn, Koch and Ziegenspeck may eventually appear of great significance in adding new and important details to the knowledge of microbic dissociation and the reproductive mechanism of bacteria; but their present interpretation and conclusions are opposed to some of the most significant facts now recognized in the bacteriophagic reaction.

Béguet's Osmotic Reaction.—The latest theory of interest is one presented recently by Béguet [31] of the institute at Algiers, and deals with assumed osmotic changes occurring in the bacterial cells in response to the nature of the colloids in their environing medium. The following experiment is cited. A culture of a sensitive ("ultrapure") Shiga dysentery strain was spread heavily over an agar surface, but in such a manner as to leave a central clear area about 1 cm in diameter. When the culture had grown for 48 hours and had "vaccinated" the medium, the bare area was inoculated with more of the original culture, and the plate was reincubated. The culture arising on this site was thin and delicate. On culturing to fresh medium, various new colony types arose. Among these was one which was limpid and hardly visible. On further cultivation this type slowly returned to the "normal" culture. There was also a colony form that was more opaque and frosted. This grew even more rapidly than the normal sensitive, gave an agglutinative form of growth in broth and was resistant to bacteriophage action. It resembled the usual secondary resistant cultures obtained from bacteriophagic action. Next, Béguet prepared a culture in P_H 7.8 broth from one of the opaque colonies and, after four days growth, added to this a loop of culture of the "ultrapure" type. After 24 hours, the medium was filtered through a Chamberland L [3] filter. The filtrate contained the Shiga bacteriophage, which could be exalted by subsequent passages. In this is found one more interesting instance of the origin of the lytic principle in the bacterial culture itself.

For explanation of these facts Béguet proposes an ingenious theory. The cell colloids of the bacteria from the opaque colonies on the semivaccinated agar have a lower osmotic pressure than the colloids in the bacteria of the "ultrapure" colonies. When one adds some of the sensitive bacteria to a filtrate of the secondary or opaque culture, the sensitive cells fix by adsorption the colloids present—just as a piece of silk (to use Béguet's example) becomes colored when placed in a liquid containing invisible traces of the dye. Since the internal osmotic pressure of the microbes is higher than that of the colloids which they have absorbed, a loss of equilibrium is produced and the organism swells like a vegetable cell in a hypotonic medium. Thus, the microbes with a pressure differing most from that of the filtrate cannot resist the swelling; they therefore burst, and their granules become scattered through the medium. This sudden dispersion constitutes lysis. Only those organisms persist and reproduce which are able to come into an equilibrium with the colloids of the filtrate; and these are the secondary, resistant cultures of d'Herelle and others.

Thus, the lytic principle, according to Béguet's view, "is regenerated by the bacterial colloids of low osmotic pressure freed in the liquid." "In the case of slight, initial differences, the phenomenon becomes arrested at the stage of swelling or of agglutination, and the culture recovers. If there is a great difference, the lysis is rapid and complete and no microbe escapes." Béguet summarized his views with the statement that the phenomenon of d'Herelle is a complex reaction in which one creates a rupture of the osmotic equilibrium existing between the germs and the absorbed colloids, or the colloids present in the medium. Some cells are destroyed by an excess of internal pressure, while others can adapt themselves to a lower pressure, thereby becoming resistant. In a second part of the same work the author brings variations of osmotic pressure into relation with the problem of variation and mutation in bacteria. This aspect, which has some points of independent interest, is considered in another section of this review.

The attempt to explain transmissible autolysis as dependent on purely physicochemical principles operative through osmotic action, or in some other way involving the cell wall or membrane, has been attractive to several investigators. Even da Costa Cruz has indulged in such notions, although he did not apply them with the thoroughness or pertinacity that characterizes the study of Béguet. But all such theories are, I believe, doubly at fault. First, they take undue liberties with our present knowledge of physical chemistry as applied to colloids in the bacterial cell, and the power of absorption from the medium. Second, they disregard absolutely the biologic aspects of the dissociative reaction, and especially those phases conditioned by the inherent hereditary mechanism, making possible the natural sequence of types in the cyclogeny of the species. One might almost come to believe, after reading such a presentation as that of Béguet, that the living microorganism is merely a toy, existing only to be played on by its colloidal environment, rather than a biologic entity possessing in itself an inherent, directive mechanism. His interpretation of the "swollen" or "inflated" cells (presumably the Pettenkofer bodies of Kuhn, the "giant coccoids" of Mellon and others) observed in numbers just preceding the lytic reaction, as resulting from a rupture of the osmotic equilibrium, can scarcely be accepted. While it must be admitted that osmotic influences may sometimes produce such pictures, one must carefully distinguish between these bodies and those which occur in the course of normal dissociation.

Wollman's Theory of the "Hereditary Factors."—In the earlier studies of Wollman [336] he introduced, in a preliminary manner, a new

view of bacteriophage action, namely, that the reaction is due to corpuscular elements which arise from the bacterial cells themselves and
are transmissible from mother cell to daughter cell, or, perhaps, from
affected cell to "normal" cell through the external medium. On these
grounds he came to relate to bacteriophage action the phenomenon which
he termed "paraheredity"—the transmission of characters from cell to
cell, not necessarily directly, but through the external environment.
This reaction, which is manifestly related to the "paragglutination" of
Kuhn and Woithe [224] and others, I shall discuss at length in the
following section; and I shall suggest, on good grounds, I believe, that
both these reactions of bacteria represent only one phase of dissociative
behavior involving particularly the serologic convergence of the R type
cultures. This type of reaction also includes the "entrainement" of
Burnet,[76] as I pointed out in my earlier paper on microbic dissociation.[168]

This theory was developed later at the hands of Wollman and Mme.
Wollman.[342, 343] In a recent paper, Wollman [336] has again sketched,
though in somewhat greater detail, the outlines of his theory, which
is now termed the hypothesis of the "hereditary factors" ("facteurs
héréditaires").

From the author's various and interesting contributions one notes
that the hereditary factors (which correspond to the corpuscles of the
bacteriophage) are conceived of as bodies which simulate those deter-
miners of heredity, the "pangens," postulated by Darwin long ago. As
may be recalled, they had their first origin in Darwin's necessity of
discovering a physical basis for the mechanism of the assumed inheritance of acquired characters. Wollman now believes that bodies of a
related nature represent the bacteriophage corpuscles; these are the
" 'facteurs' héréditaires." These factors (which naturally possess a
physical basis), having once arisen in the bacterial cell, can acquire a
certain autonomy and can produce in "normal" cells the modifications
of which they are the bearers.

Thus, Wollman's hypothesis attempts a sort of reconciliation between
the "hereditary" aspects of the theory of Bordet and Ciuca and the
"transmissible" aspects of the theory of d'Herelle. In order to accomplish this reconciliation the pangens of Darwin (now "hereditary
factors") are attributed with the ability of transcending the limits of the
cell in which they arose and of existing in a free state, endowed with
all the stability which one has learned characterizes the bacteriophage.
Encountering the "normal" cell, they are able to impress on it the
disturbance of equilibrium between the anabolic and katabolic functions,

such as characterized the cell from which they arose. Here, then, one sees the delineation of an extracellular, physical mechanism capable of transmitting to other and unrelated cells the characteristics acquired by the cell first influenced. Indeed, the author considers the possibility that this conception might be extended to cover other pathologic reactions in the cells of higher forms, such as the sarcoma of Rous and the mosaic diseases of plants.

However speculative these views of Wollman may appear, it seems to me that, in attempting to escape the difficulties of the diastatic and parasitic theories, and to establish the existence of a cellular element, not strictly autonomous, but, nevertheless, related biologically and genetically to the bacterial cell and transmitted through the external medium to other and "normal" cells, this author has made a distinct advance toward a more logical interpretation of bacteriophage action. To most biologists, if not to bacteriologists, the chief difficulty in Wollman's theory will undoubtedly be the conception of the transmission of acquired, hereditary characters through the external environment. And along with this, as Wollman himself has suspected, will occur the difficulty of accepting the existence of "intracellular 'factors'" capable of acquiring a certain degree of autonomy and even remaining intact in the external medium." Moreover, many biologists will hesitate to accept the resurrection of Darwin's pangens in a form which brings them again into relation with a mechanism for the inheritance of acquired characters—a biologic conception to which Wollman appears to subscribe. As will be noted in a later section, if, for Wollman's "hereditary factors," one substitutes certain accessory, filtrable elements arising from bacteria in the course of normal reproduction, and possessing, perhaps, some sort of fecundating influence, then Wollman's hypothesis approaches closely my own theory of bacteriophage action, which will be developed in detail in a subsequent section of the present work.

The Theory of Otto and Munter.—In concluding my consideration of these various theories of the nature and origin of the bacteriophage, the theory of Otto and Munter remains to be mentioned. This has not been regarded by bacteriologists as a noteworthy contribution, since it involved mainly the lysin hypothesis of Bordet and others, with only certain apparently insignificant modifications of some of the details. But, however far afield it may go in suggesting the identity of the bacteriophage with minute bacterial particles possessing enzymatic power, this theory has the great merit of introducing into the arena of argument an important point which has been lacking in all other considerations; and which, although recently vigorously attacked by Sonnenschein,[315] has, I believe, a fundamental place in any explanation of the bacteriophage reaction.

I refer particularly to the view of Otto and Munter that the bacteriophage cannot be obtained from a given culture at any time, but that there exist certain moments ("Momenten") in the development of the culture when it is in the most favorable state for the spontaneous generation of the lytic agent. I do not wish to develop this point in detail at present, but leave the matter with the suggestion that the "Momenten" of these authors may, in reality, represent the entrance of the cells of the culture into that cyclostage in which the lytic corpuscles are liberated from the bacteria. There can now be little doubt that the "Momenten" are concerned with the evolution of certain cell structures perhaps identical with the Pettenkofer A bodies of Kuhn. The significance of these points will become clearer in the following section.

The Dual Nature of the Lytic Agent.—In practically all theories of the nature of the bacteriophage there exists one objection that stands out beyond all others, and one of which neither d'Herelle nor Bordet have taken sufficient account. This is the circumstance that all present major hypotheses regard the bacteriophage as represented either by homologous corpuscular units, or by an assumedly homogeneous enzyme-like substance. For our present considerations, the latter conception may be dismissed since the general view of the enzyme nature of the bacteriophage has a sufficient number of serious objections. But, regarding those explanations of the nature of the bacteriophage which interpret the agent as comprising some sort of physical units, something more needs be said at this point. Our present consideration may be brief, however, since the experimental data and observations which support a quite different view have already been presented in earlier pages of this section, and in the paragraphs dealing with variations in the size of the lytic plaques; and other points bearing on the problem will be offered in subsequent sections.

One of the greatest defects in d'Herelle's exposition of the nature of the bacteriophage is found in his failure to observe that the behavior of the lytic units (his virus micellae), active on any given species, enables us to arrange them, not in a continuous series of variations, but in a *discontinuous* series. The complete bacteriophage is therefore not homogeneous but heterogeneous; the units comprising it are not all the same, but represent distinct groups. This heterogeneity was early observed by Bail[26] and by Watanabe,[328] and these investigators attempted to depict the degree of this heterogeneity by the postulation of three "elementary species" of bacteriophage. This heterogeneity was also recognized by Gratia [150] and its degree was regarded by him as being much greater (involving the possibility of a different bacteriophage for each and every different variant of the acknowledgedly pleomorphic colon species) although his actual experiments demonstrated the existence of only two types of bacteriophage unit—one producing large areas, the other producing small. Most recently this same heterogeneity has been noted by Kline who has recognized only the dual mechanism. The studies which my students and myself have conducted on this problem, involving investigations on B. dysenteriae Shiga, B. typhosus, B. paratyphosus A, B. paratyphosus B, B. proteus and M. albus, have led me to the conclusion that the actual degree of heterogeneity observable among the lytic units of a single species does not transcend the limits of a dual system; that, for each bacterial species that has been adequately studied, there exist only two forms of the lytic agent, namely the a and the β units, producing respectively the large and the small lytic areas. Wherever a third form of a given principle seems to make its appearance (as indicated by plaques of a different shape or size) it is only a minor variation of either the a or the β types. At the present time there is no evidence that there exists a type intermediate between the a and the β principles. If, therefore, the complete bacteriophage for a given culture is regarded as comprising certain corpuscular units, whether of a virus-like or some other nature, it would appear that these units are of two sorts only. Before going further with this discussion it may be well to bring together the evidence that there actually exist such diverse bacteriophagic units, as opposed to the view that the results observed are to be interpreted as variable effects from the action of a homogeneous system.

If the only evidence to support the conclusion that the bacteriophage in its entirety is made up of two types of lytic unit were the observation that there exist, for any single bacterial culture, lytic areas of two different sizes—the large and the small—this view might well be doubted; for it is conceivable that a lytic filtrate composed of homologous units might reveal areas of different size, depending on the form of culture or the type of cell in a heterogeneous bacterial population coming under the influence of a homogeneous bacteriophagic stimulus. But, fortunately, there are other grounds to support this conclusion.

In the first place, it is apparent that the two forms of area differ in other characteristics than that of size. The large areas are more likely to be diffuse while the small are clear-cut and sharply differentiated.

The margins of the large areas are usually indistinct and hazy while the margins of the small areas are clean.

The two principles differ also in respect to the cyclostage of the culture that is most vigorously attacked. The α units influence either the S type culture or certain secondary cultures that have proved resistant to the β principle. The β units, on the other hand, influence either the S type culture or certain culture forms that have proved resistant to the action of the α principle. Neither the α nor the β units, however, so far as has been ascertained, are able to influence appreciably cultures of the intermediate or O type.* Moreover, there is some slight evidence that, when a certain strain of bacteriophage possesses the ability to attack heterologous, but at the same time somewhat related, bacterial species, it is the β principle, rather than the α, that accomplishes this result. Here we may cite the influence of the paratyphoid β units on the typhoid S type culture; but the failure of the paratyphoid α units to produce lytic effects in the typhoid S substratum.

In the second place, additional evidence supporting the distinct nature of the α and the β principles is derived from the differences in the secondary, resistant cultures which arise after, or in the course of, the lytic action of these respective principles. This phase of the problem requires much more study but the facts thus far obtained indicate that the action of the α principle on S type culture permits the survival, or generation, of a rough R form of culture which seems to differ little, if at all, from R type cultures obtained under the influence of many other known incitants to the dissociative reaction, such as aging, pancreatin, lithium chloride, etc. The secondary, resistant cultures arising from the action of the β principle are, on the other hand, usually not rough forms but smooth. In what manner they differ from the "normal," smooth S type culture is not yet known; but they often resemble the S type culture in respect to the point that they may again become susceptible to the action of the α principle, to which the rough resistants are commonly immune.

Next should be mentioned the difference in the mode of action of the antilytic serums prepared from the α and β lytic filtrates. These differences, first pointed out by Bail, by Bail and Watanabe, and by Watanabe were confirmed by Gratia in 1923; and, though lacking in some impor-

---

* Nor, from some quite recent studies performed by my students, Klimek and Delves, on Shiga dysentery cultures, does the α principle act on the coccus-like gonidial form of the Shiga bacillus, which we have succeeded in stabilizing for a considerable number of generations. This culture type is composed of elements which often appear in a streptococcus grouping and which vary in size from 0.2μ or less to 0.5μ. On agar the colonies appear as minute bluish points from 0.1 to 0.3 mm. in diameter. In broth they give a coarsely granular growth whcih leaves the medium perfectly clear, or grow in the form of a faint precipitate at the bottom of the tube. This new form of culture will be described in greater detail at a later date. It is one of the two gonidial forms characteristic of the Shiga culture.

tant details, the study of a paratyphoid B bacteriophage by Eugenia Dabney and myself [169a] shows the specific inhibiting effect of a serum immune to the α principle when permitted to act on its homologous principle in contact with the sensitive culture; also the failure of this serum to inhibit the β principle. Experimental work of this sort is difficult because of the danger that the serum may influence the substratum culture (making it resistant) as well as the bacteriophage. But it seems possible tentatively to explain the results obtained by Bail, Watanabe, Gratia and ourselves on the basis of a distinct difference in the antigenic structure of the α and the β units, respectively.

Finally, and as the most significant difference which exists between these two principles, must be mentioned the dissimilarity in heat-stability. Since this aspect of the question has already been discussed, it is now sufficient to recall attention to the fact that, at least in certain lytic agents active on members of the colon-typhoid-paratyphoid-dysentery group, there exists a difference of about 10 degrees C. between the points of thermal inactivation of the α and β principles—the former being rendered inert by an exposure of one-half hour to a temperature of 62 to 63 C., while the latter requires for inactivation a similar exposure to a temperature of about 75 C. Incidentally, it might be pointed out that this difference in the point of thermal inactivation of the two principles may underlie the discrepancies apparent in the literature bearing on the subject of the point of thermal inactivation of the bacteriophage. Here it may be recalled that different investigators have reported the effective temperature at various points between 62 and 74 C. In addition, these observations may explain the circumstance that heating a lytic filtrate at some intermediate point—such, for example, as 68 C.—serves to weaken its lytic action, but may not render the filtrate inert. In such a case the weakening action would be due to the elimination of the α units.

All evidence considered, it is difficult to controvert the fact that the bacteriophage, in its totality, possesses not a single but a dual personality; and that the two principles concerned are represented by two antigenically distinct, but in a measure functionally-reciprocal, units. These facts are impossible of explanation on the grounds of d'Herelle's filtrable virus theory; and they can be explained no more readily by appeal to Bordet's (or any other) enzymatic conception of the nature of the lytic agent. A possible interpretation of these peculiar reactions on the basis of the dissociative phenomenon will be presented in a subsequent section.

Conclusions.—The list of references presented on a preceding page, and others that might be added, readily indicate that the majority of recent workers with the bacteriophage have adopted Bordet's view of the agent as an autolysin, implying, of course, its nonliving nature. Indeed, Bordet stated two years or more ago that d'Herelle would soon be left as the sole supporter of his virus theory. This is undoubtedly true, for it is seldom, now, that one encounters further reports of a sustaining nature. In this regard, however, it is a curious and unfortunate circumstance that the loss of prestige of d'Herelle's view has not followed the actual disproof of any of his splendid array of facts, nor has it followed the setting up of counterproofs. It seems to have been determined largely by that poorly defined, and scarcely scientific, feeling of "general improbability."

In view of the increasing ascendancy of Bordet's conception, the question naturally arises, how does it happen that this view has gained so many adherents when it is realized that none of its supporters, nor even Bordet himself, has been able adequately to explain, on this hypothesis, the most characteristic of bacteriophagic phenomena—the production of plaques—as well as several other reactions of much significance? The explanation seems to be that most of the supporters of the Bordet hypothesis apparently have voluntarily closed their eyes to the necessity of explaining this important feature; or have abandoned the task to their leader and to the still inadequate support that he has been able to adduce. Here, then, in the case of Bordet's theory also, its ready acceptance by the majority appears to rest on the rather insecure basis of a feeling of "general probability." This circumstance, in the minds of some investigators at least, has proved a great source of weakness in the autolysin theory, and has, in reality, added some support to d'Herelle's conception—as a result of letting certain interpretations and conclusions go "by default," as it were. It stands to reason that the explanation of the bacteriophage problem as a whole will not be available so long as only certain aspects of the reaction are explained. At present, it must be frankly admitted that d'Herelle's theory is the one that explains most satisfactorily the greatest number of observed facts. Moreover, the theory is so strongly fortified that it can never be disproved by resort to juggling and to reinterpreting the facts that he has presented, for he has contributed the only possible interpretation of this body of facts as they now stand. In other words, the disproof of d'Herelle's view, as also of the theory of Bordet and Ciuca, depends entirely on the accumulation and analysis of new facts.

But where, one may pertinently inquire, can such facts, possessing any fresh significance for this intricate problem, be found? The field of inquiry in all its aspects seems to have been worked over thoroughly, and all other avenues of approach to have been closed. In a previous publication I have already expressed my conviction that these facts, so urgently needed, are to be discovered in a new but highly significant branch of bacteriologic study which is just beginning to emerge from a state of random observation and to assume a concrete form, namely, the field of microbic dissociation. Toward this field, though often by devious paths, and sometimes unconsciously, many investigators are already beginning to turn their steps.

10. THE RELATION OF TRANSMISSIBLE AUTOLYSIS TO THE PHENOMENON OF MICROBIC DISSOCIATION

In approaching the theory of bacteriophage action which I desire later to propose it is necessary to discuss, at this point, an aspect of the reaction which has been left quite unconsidered by d'Herelle, and almost equally so by Bordet, but regarding which, several obscure, but at the same time significant, references have appeared in the literature of the past few years. I refer to the phenomenon of microbic dissociation, a general and critical review of which I have presented in an earlier paper (1927).[168] As I have there shown, this aspect of bacteriology embraces phenomena which penetrate every field of bacteriologic study; and the problem of the bacteriophage no less than others. It is, moreover, a subject in which the older tenets of monomorphism, which have dictated the views of d'Herelle and, to a lesser degree, of Bordet, cannot find a place. In the light of the present knowledge of dissociative reactions, one will readily see that the conceptions of these workers have followed explanatory trends that are little in harmony with certain fundamental facts concerning bacterial behavior as we are now coming to view them.

In introducing this new aspect of the subject, in which will be found some of those facts needed to clarify the views relating to the fundamental nature of bacteriophage action, I may, at the outset, mention an interesting and perhaps significant circumstance; namely, that there have been developing, particularly during the last decade, three aspects of bacteriology which are likely to mark this period as one of almost unprecedented significance in the history of the science. These are: (1) the discovery of the bacteriophage; (2) the clear recognition of the existence of filtrable stages in the development of bacteria; and (3) the threatened abolition of monomorphism as a dominating point of view in the study of bacteria, accompanied by the substitution of a conception of a rational plurimorphism, based on bacterial cyclogeny.

Each of these varied aspects has had, up to the present, an independent development; and each in turn has seemed to validate many and varied, earlier observations in its respective field. But the point which I wish to make clear at present is that, as has so often happened in the history of biology, certain unmistakable evidence is accumulating to show that these three aspects of bacteriologic study are converging, from one angle or another, on a single and most fundamental problem,— the reproductive mechanism of bacteria. This statement cannot be proved at once, or in a few words; but to any critical observer who has made himself familiar, to an extent, with these three fields of study, the signs are ominous.

In the present limited review it is not my intention to attempt to establish all the correlations that might be made between these three fields of investigation, or to indicate the important bearings that each may have on the others; and much less to consider the interesting details of those recent studies that afford the basis for the conception of bacterial cyclogeny and the sequence of developmental forms which it embraces. In the present section all that I can hope to accomplish is to draw attention to certain important relations between two of these fields of study,—the phenomenon of the bacteriophage and the phenomenon of microbic dissociation.

REFERENCES *

Arkwright[6, 9] (description of the S and R types).

Eastwood[112] (relation of bacteriophage action to phenomena of bacterial variation).

Gratia[148, 150] (dissociating influence of bacteriophage).

Gratia[155, 156] (relation between bacteriophage action and the S and R types).

Breinl and Fischer[30] (comparison of the effects of aging and of bacteriophage action in producing variants).

Breinl and Hoder[31] (similarity of the "mutants," whether produced by aging, by high temperature or by lytic agent).

Hoder[197] (bacteriophage action stands in important relation to bacterial mutation).

Enderlein[117] (the basal facts of bacterial cyclogeny, mainly from the microscopic features of the cells).

Wollman and Wollman[342, 343] (bacteriophage action and para-hereditary transmission of characters).

Otto and Sukennikowa[271] (relation of bacteriophage action to paragglutination).

* In this reference list are presented only those citations dealing with the correlation of bacteriophage action and bacterial "variation" or "mutation," in such a manner as to suggest that the authors observed some distinct, though unexplained, relation between these two phenomena. Other references bearing less closely on the problem are to be found in the general bibliography at the end of this review.

Otto and Munter [204] (bacteriophage generated by bacteria themselves, but only perhaps at certain [physiologic] "Momenten" in the life of the culture).

Kauffman [211] (relation of variants observed in "Keimumwandlung" and in bacteriophage action).

Kuhn [221, 222] (relation of bacteriophage action to Pettenkofer bodies).

Wollman [336] (description of dissociative variants).

Grumbach and Dimtza [359] (related to the "mutation" phenomena of Neisser and Massini; description of "ephemeral modifications").

Koch and Ziegenspeck [215] (further evidence of the relation of the bacteriophage to the Pettenkofer bodies).

Hadley [166] (review of the phenomenon of microbic dissociation; its relation to bacteriophage action).

DISCUSSION

Correlation of Bacteriophagic Reactions with Variation Phenomena in Bacteria.—After the important contribution of Arkright in 1921, making clear many of the points of differentiation between the S and the R types of bacteria, Eastwood, in an article entitled, "Bacterial variation and transmissible autolysis," suggested that these "aberrant" forms might be just as normal as others with which the bacteriologist usually has to deal. He pointed out that the bacteriophage also produced such variations, and that it should not be regarded merely as an agent which destroys bacteria. He also recognized the fact that some of the variants produced by the bacteriophage resembled closely some of those described before the bacteriophage was known. Further, that the bacteriophage may not be a new discovery but "simply a revival   .   .   . of old ideas and observations about the lytic properties of bacterial enzymes." Eastwood believed that the lytic phenomena might be closely associated with other observed facts concerning bacterial variation when lysis is not observed, and that the two might be a part of the same general problem. Although this author apparently leaned toward the enzyme nature of the bacteriophage, attributing the reaction to a rupture in the equilibrium existing in the anabolic-katabolic balance in the cells, his other views are of great importance in that they approach the point of view which I wish to develop in the present section.

Experimental evidence bearing on this general conception of relation between bacteriophagic reactions and variation was presented in 1923 by Breinl and Fischer. These workers, without recognizing the dissociative reaction as such, clearly noted dissociative behavior in old cultures of B. paratyphosus B and B. cholerae suis, from which certain variants were derived. They stated that the production of variants by the bacteriophage is at least as marked as can be obtained from old cultures aged in a liquid medium. This last, it may be noted, has been for many years a favorite and reliable method of obtaining the dissociated bacterial types.

In 1925, Breinl and Hoder reported on the "mutants" of B. paratyphosus B, produced by the bacteriophage, and showed by serologic tests
the type of antigen (large flaking or small flaking) present in the chief
mutants. Their conclusion from this work is of special interest. The
authors believed that the actual dissolution and destruction of the
bacteria in the bacteriophage reaction is only a special case, and certainly
an extreme one; for a greater part of the microbes experience a more
or less far-reaching modification in which form they are differentiated
from the original type culturally, serologically and in resistance to the
lytic action. In most of their derived forms, stable varieties were
not found. Many of the modified forms were not "end-points" but
intermediate stages which, under longer influence of the lytic principle,
were further transformed into one or two fairly uniform types. The
longer the bacteriophage "worked" on the cultures, the more uniform
became the variants. Often, all the varieties were eventually transformed into a single resistant type. It is important to note that these
changes were analogous to those observed in cultures aged in broth and,
as is known, undergoing normal microbic dissociation. Thus, the
conclusion was reached that the bacteriophage reaction stands in close
relation to bacterial "mutation."

In 1925, also, similar experiments were reported by Hoder under
the title, "Regarding the Relation Between the Bacteriophage and
Bacterial Mutation." These dealt mainly with members of the colon-
typhoid-dysentery group. In the explanation of his results, however,
Hoder (as was also the case of the investigators just previously mentioned) turned to Bail's "splitter" theory, rather than to the explanations
of the dissociative reaction which I have elsewhere advanced. This
circumstance, however, does not detract from the value of his facts, for
these easily can receive another interpretation.

Wollman,[335] in a general review of the bacteriophage in 1925, as
also in his comprehensive paper published in 1927,[336] concluded that
the virus theory was slightly probable, and that the same was true
of the diastatic theories (Bordet's and others). Lysis, he believed,
cannot be related to known ferments. Moreover, he assumed that the
antigenic properties of the bacteriophage showed that it does not concern normal bacterial ferments. While not accepting in toto the theory
of Bordet and Ciuca, he believed that these workers had introduced a
conception of importance, namely, that of heredity, an aspect of the
subject which Wollman develops in a new manner. As mentioned in
the preceding section, he considers whether the lytic agent could be

the result of a variation of which the determining element would be corpuscular and transmissible, "perhaps from mother cell to daughter cell, perhaps from affected cell to normal cell through the external medium." Into possible relation with this view he brings the pangens of Darwin; but, with this exception, which enters the field of speculative zoology, it clearly appears that Wollman caught the glimpse of biologic factors of significance, developed in greater detail by Wollman and Mme. Wollman in 1925 and 1926 under the heading of "transmission parahéréditaire" of characters among bacteria; also by Wollman [336] in 1927. In the first of these studies the Wollmans recognize the possible relation of "paraheredity" to the "paragglutination" of Kuhn and Woithe (1909). Before proceeding further, it therefore becomes desirable to examine briefly the latter phenomenon.

Paraheredity and Paragglutination.—The notion of paragglutination was first developed by Kuhn and Woithe [224] in 1909. By this term the authors referred to a phenomenon in which a specific immune serum from a dysentery patient, for example (dysentery Y in the original case), agglutinated heterologous species, such as B coli, coming from the same dysentery patient. In explaining these reactions the authors suggested that the heterologous germs became endowed with receptors for the Y agglutinins, and in 1911 Kuhn, with Gildemeister and Woithe,[223a] seemed to demonstrate that coli germs could be similarly endowed experimentally with receptors for dysentery agglutinins. Rabbits highly immunized to dysentery Y were injected with B. coli culture. The coli strain when recovered was so modified that it agglutinated in dysentery serum in 1:200 dilution, while the reaction had previously been negative. Unfortunately, however, the same modified coli culture was agglutinated by normal rabbit serum. In 1916, Kuhn and Ebeling [223] showed that paragglutination occurred in B. coli as a result of growing it on an agar medium in which broth cultures of other bacteria had been incorporated. The paragglutinative reaction was manifested after two passages on such medium, but was increased after three, and reached its highest point after five; then it dropped off again.

Otto and Munter,[269] in 1921, suggested that the bacteria undergoing this reaction might have some relation to the bacteriophage lysin. In 1923, the subject was again taken up by Otto and Sukennikowa.[271] These authors studied the effect of a Flexner bacteriophage on the paragglutination of B. coli in Flexner antiserums. By the growth of several strains of coli in Flexner lytic filtrate some came to present paragglutinative properties. This result did not occur, however, when a Shiga or a dysentery Y lytic principle was used. The property was lost with further cultivation. Since similar results, though exhibiting weaker reactions, could be obtained by placing B. coli in contact with heated Flexner cultures, the authors concluded that paragglutination was not necessarily bound up with the bacteriophagic lysins, although these favored the production of paragglutinating receptors.

As closely related to, if not identical with, the phenomenon just described, may be mentioned the observations of Bachmann and de la Barrera [22] on the serologic behavior of "mutants" of B. paratyphosus A under the influence of the lytic agent. In this case, some of the mutants manifested a marked agglutination in serum immune to B. typhosus. After many transfers there was obtained from one of the cultures an immune serum which agglutinated the mutant in high dilutions, also B. typhosus. But this serum failed to agglutinate the original paratyphoid culture. Here manifestly is a modified culture type which arose under the lytic stimulus, and which takes on new and heterologous antigenic characters.

I now come again to the observations of Wollman and Mme. Wollman.[342, 343] In 1926, they reported further on parahereditary transmission of characters in B. coli, suggesting that this phenomenon might be related to paragglutination. They mixed with a culture of B. coli a culture of B. paratyphosus B and made four passages. They next separated coli by plating and then made seven further passages to purify the strain. Two strains were then selected and tested, together with the original coli antigen, against serum immune to B. paratyphosus. The original coli culture was agglutinated 1:80; the "modified" coli culture at from 1:640 to 1:1280. Five further passages of the modified strains did not produce further change in them. When cultures of B. coli and B. typhosus were mixed in a similar manner, modification of coli was not observed. The Wollmans call attention to the fact that E. Burnet succeeded in causing the "influence" to pass filters, and even collodion membranes, in the case of M. melitensis. The probable nature of Burnet's reaction I have considered in detail in an earlier paper.[168]

To those who have followed the recent trend of studies dealing with bacterial convergence resulting (among other effects) in the serologic cosmopolitanism of Schutze,[302] and have caught the significance of the more recent work of F. M. Burnet,[78] dealing with the parallel trends existing between serologic and bacteriophagic reactions determined by the heat stable antigens of bacteria, the studies on the phenomenon of paragglutination will be recognized at once as a special case of serologic convergence. It can be tentatively assumed that such influences as those just reported, brought to bear on the S type coli culture, serve to cause a transformation characterized by an increase in the content of the O or the R heat stable antigens. Owing to this modification the culture gains a marked antigenic heterogeneity by virtue of which it is agglutinable by serums which previously affected it slightly, if at all. This is no more than has already been shown by the illuminating studies of Schutze, Bruce-White,[62] Goyle and Balteanu for the O or R types of culture produced by forced dissociation and without implication of the bacteriophagic reaction; also by F. M. Burnet [78] for the R type antigens in relation to bacteriophagic action. In all these cases a considerable degree of antigenic heterogeneity was developed, and I do not see any reason for assuming that the circumstances last mentioned are essentially different from those reported by the Wollmans (bacterial association affecting serologic convergence) or from those of Otto and Sukennikowa [271] (influence of bacteriophage in accomplishing the same end). This is not, of course, an actual explanation of the phenomenon, for it does not tell us the nature of the common mechanism present in aging cultures, bacterial filtrates, microbic associations, chemical stimulation or bacteriophagic action, or its rôle in the production of the observed transformations in the culture type.

I have presented the experimental data bearing on this subject in some detail merely for the purpose of indicating that reactions attributable to the bacteriophage in the phenomenon of paragglutination are not necessarily different from those determined by normal dissociation occurring without the intervention of the lytic filtrates. In all the cases mentioned it is safe to assume that the antigenic analysis (or synthesis) occurring under the cloak of the dissociative reactions, however instigated, yields the results characteristic of the phenomenon of Kuhn and Woithe; and also of the "paraheredity" of the Wollmans. Both reactions are intimately—one may even say inextricably—concerned with microbic dissociation and, pari passu, with the bacteriophagic reaction. They are both, however, the results of the process and not the cause.

Another instance in which variations and mutations have been brought prominently into relation with the bacteriophagic reaction is found in the recent work of Béguet.[31] It is to be regretted, however, that, in attempting to explain the interesting phenomena that he observed, he was led into the realm of physicochemical reactions involving particularly osmotic phenomena. The facts presented are none the less valuable, however, from the dissociation point of view. For one thing, he adds to the evidence already gained that, before lysis, many of the organisms become swollen or "inflated," and that it is these forms especially that disappear in lysis. Thus, his observations confirm those of d'Herelle, Wollman, Kuhn, Larkum and some others, although his interpretation is entirely different. Moreover, he points out a type of culture which he terms the "type à colloïdes condensés," characterized by opaque, dry colonies, and giving a sedimentary growth in broth. At the same time, it was less actively agglutinated in specific serum and less virulent, and possessed greater vitality. This is manifestly an R type culture. Opposed to this culture type was his "type à colloïdes dispersés," which gave smooth, transparent colonies, showed a tendency to superficial growth in broth, agglutinated readily in specific serum, was more virulent and possessed less vitality. This is manifestly the S type culture. These forms, he stated, were usually transitory; but he succeeded in obtaining "veritable mutations" that were persistent on any medium. Many other curious aspects of bacterial behavior are, according to Béguet, interpretable on the grounds of differences in osmotic pressures; namely, the filtrable forms of B. tuberculosis, B. dysenteriae and B. pestis; the modified characters of the B.C.G. cultures of Calmette; the lowered virulence and agglutinability of old cultures, and such curious transformations as that of the pneumococcus into streptococcus, alleged by several investigators. In these and other curious reactions among bacteria he assumes that osmotic pressure reactions hold the controlling influence.

From this it appears that Béguet has viewed some of the striking pictures of dissociative changes in bacterial cultures, but without catching the significance of the actual mechanism underlying the reactions. For him the bacterial cell can merely respond to environmental stimuli; it cannot itself initiate reactions nor determine, through innate powers of variability, the trend of its own cyclogenic evolution. That environment is strongly instrumental in modifying the bacterial type no one can deny; but it will also be seen that there lies in the bacterial protoplasm a directive influence of great significance. The biologic fallacies inherent in Béguet's conception will become clearer in a later section.

Transmissible Autolysis as a Dissociative Reaction.—Aside from experiments in my own laboratory, the works cited at the beginning of this section comprise the chief references in which the authors seek to relate the bacteriophage reaction with bacterial variations or mutations.

None of these authors, however, has approached the point of view of the possible relation of the observed transformations in cell or culture type to the phenomenon of microbic dissociation in the sense in which I have employed this term in my earlier review.

In order to conserve space, I shall refer here merely to some of the broader points of resemblance between the bacteriophage reaction and microbic dissociation, referring the reader to my earlier work for the details, as well as for the exact references.

With reference to the matter of incidence of these two groups of phenomena, I [168a] have already pointed out that the bacteriophage reaction has been observed only in those bacterial species also known to present features of the dissociative reaction; and this is particularly clear among the members of the colon-typhoid-dysentery group, the various members of which have served as the experimental material on which the greater part of the research on the bacteriophage has been based.

Regarding the production of variants or "mutations," I have conclusively shown that, contrary to the conclusions of d'Herelle,[188, 193] these cultural modifications are not the products of bacteriophage action alone, at least not in the sense in which the term is employed by him, but appear in many bacterial cultures when submitted to the influence of a dissociation provoking environment. Grumbach and Dimtza [359] have recently reached a similar conclusion. There is an almost perfect analogy between the chief culture types of the dissociation series and those concerned in the lytic reaction. The smooth S type culture, most sensitive to the dissociation provoking influences, is identical with the form most sensitive to the lytic agent. The O culture type of the intermediates in the dissociation shows many points of resemblance with the transitionals revealed in the bacteriophage reaction, and the R types from dissociation have a close counterpart in the culture forms that are more resistant, or quite resistant, at least, to the a bacteriophage.

With reference to the occurrence of the two reactions in culture mediums I have pointed out certain resemblances. Neither reaction will occur in a medium in which the microbes cannot multiply. In salt solution suspensions the bacterial cells undergo neither dissociation nor transmissible autolysis. In this connection, it has always seemed strange to me that, if the bacteriophage is a filtrable virus, it cannot act on such suspended cells. The reaction of the medium which most favors bacteriophagic reaction (according to d'Herelle, P_H 7.8) is also most favorable for dissociation.

With reference to the permanence of the new forms, the resistant types arising from bacteriophage action are usually not susceptible to further lysis by the same principle, and may retain their new characteristics for a long time—sometimes, according to d'Herelle, they represent permanent modifications. In time, however, the majority of them seem to revert to the original form of culture. According to Bronfenbrenner and Korb,[59] the resistant form of B. pestis caviae characterized by the large, rough colonies is much more stable than the resistant form characterized by the small, smooth colonies and yielding the flocculent growth in broth. But it is also known that the rough culture form arising from normal dissociation maintains its new characters for a long period, although it is commonly retransformed to the smooth sensitive form, even after months of cultivation. On the other hand, some investigators have reported that the rough type sometimes represents a permanent modification, and such cultures have commonly been regarded as mutants.

With reference to the production of filtrable forms of bacteria [168b] these have been observed, not only in cultures that have undergone transmissible autolysis, but in normal, sensitive cultures in which the bacteriophage reaction has not been noted at the time. I cannot yet say definitely that they have been detected in cultures recognized as existing in the active state of normal dissociation, but it seems probable that this will eventually be the case.*

Regarding proliferative growth,[168c] it has been shown that this occurs equally in cultures approaching the lytic threshold (or "critical period," as termed by Bordet) in the bacteriophagic reaction, and in cultures in the process of normal dissociation. In both there arise, about this time, the specialized cell forms which Hort [352] regarded as announcing what he termed the "reproductive explosion."

Numerous experiments have indicated that the so-called lysogenic cultures [168d] are readily produced by the action of a weak lytic filtrate on a sensitive culture. But it has also been noted that cultures marked by lysogenic activity are observed in the dissociative reaction, and that these cultures may sometimes retain their characteristics for a considerable time. On the other hand, such cultures sometimes disappear rapidly from solid culture mediums, and are then referred to as "suicide" cultures. They also may leave resistant colonies in their path.

---

* Since writing the above, I have, with my students, Edna Delves and Klimek, obtained a filtrable stage in the cyclogeny of B. dysenteriae Shiga as a result of dissociation forced by the influence of lithium chloride and pancreatin (Squibb's). In another case the filtrable form of Shiga arose in the tenth generation passage through normal infusion broth. With my students, Frohman and Ratner, I have studied a similar cyclostage in B. typhosus, dissociated by lithium chloride.

With reference to the sort of modification produced in the secondary cultures arising from lysis or from dissociation, it may be added that these commonly resemble one another in loss of virulence and in the acquisition of ready phagocytability. I have already [168e] presented many instances supporting this point. Certain of the resistant cultures from each source resemble each other further in their agglutinative growth in broth. This is especially the case in the resistant forms characterized by the large, rough colonies, and sometimes still containing remnants of the intermediate type cells. Some resistant cultures, however, give homogeneous clouding.

In connection with these forms and their greater resistance to the original a bacteriophage, another point is of interest. This concerns the observation that only young, sensitive cultures are most susceptible to the principle; and that, when such cultures are aged, or are acted on by certain incitants to dissociation, they become more resistant. How can it be explained that only the cells of the smooth S type are most susceptible; and that their transformation into a subsequent cyclostage determines partial or complete resistance to the same bacteriophage strain? Further—what is the actual significance of the "immune," resistant cultures of d'Herelle?

The fact that certain chemical substances applied to sensitive cultures contribute to rendering them resistant has been explained by d'Herelle by the circumstances that such cultures have been "modified"; only "normal" cultures are sensitive to the bacteriophage. At no time, however, does he draw a parallel between the cultures that have become so "modified" and cultures that have become resistant through the spontaneous or enforced dissociative reaction, or under bacteriophagic influence. In fact, the phenomenon of microbic dissociation has escaped notice throughout all of d'Herelle's work. Yet it is known that the trend of culture modification occurring in microbic dissociation parallels the trend of modification in transmissible autolysis. And it is known that the end-results of the transformation are similar. The parallelism of these lines of culture modification naturally suggests, but does not necessarily prove, a similar causative mechanism. In any case, one can observe that chemical substances in great variety, as well as certain environmental conditions, can produce from "normal" culture, forms that are more resistant to the bacteriophage, especially the a units; and one also sees that the action of the a bacteriophage on the same "normal" S type culture can produce similar resistant forms.

One is, thus, apparently confronted, in the same bacterial species, with at least two kinds of resistant culture: one that has developed either spontaneously, or under the influence of certain incitants; the other which has developed under bacteriophagic influence (a units) alone. And these two kinds of culture are remarkably alike in many of their characteristics which are not related to their reaction to the lytic influence.

D'Herelle, however, has pointed out that the resistance of a culture to a bacteriophage that is able to attack other forms of the same culture is always to be taken as evidence that the resistant culture has acquired an immunity to the ultraparasite, either recently, or at some time in its past history. How, then, shall these facts be interpreted? Shall one form of resistance (from dissociation) be called natural immunity, or "mutation-immunity" and the other form of resistance (from bacterio-phagic influence) acquired immunity? Or, shall it be concluded that the R dissociate has in some way acquired an immunity without ever having come into contact with the ultravirus? Further, there is the problem of how to differentiate between the two resistant forms of cul-ture. In case one possesses natural immunity and the other acquired immunity, one might expect that in the former the resistance would be indefinitely retained, while in the latter it would eventually be lost. But the fact is that resistance is lost, sooner or later, in both forms, although a few doubtful exceptions have been noted. All evidence considered, it seems to me that there are not any grounds for accepting d'Herelle's highly elaborated theory of resistance to bacteriophagic action as an immunity acquired by the microbes of the resistant culture.

If this is true, however, how shall the origin of the resistant types, whether arising from normal dissociation or from lytic action, be explained? Further, what is their actual significance in the life of the culture or the species?

From the evidence already presented in this and in earlier publi-cations, it seems clear that resistance to the bacteriophage of a certain type is the attribute of certain stages in the cyclogeny of the bacterial species. In these definite stages the cells as a whole are not susceptible to the advances of the lytic corpuscles; or at least the entire culture cannot be precipitated into the acute reaction. Left to itself for a time, or brought under the influence of certain foreign substances, or environ-mental conditions (there is now definite knowledge of many, which can be employed for the purpose), the culture becomes transformed along the normal cyclogenic path into stages in which it is refractory to lysis by the same bacteriophage strain; and, if it is patiently watched or encouraged by certain known means, it will return again to the state of sensitivity.* Resistance to bacteriophage action is, therefore, the normal characteristic of certain cyclostages in the developmental history of the species; or, to put the matter in another light, the bacterial species is highly susceptible to a certain lytic agent only when it has entered a definite cyclostage. It has already been learned from studies on microbic dissociation that, in the cyclogeny of certain pathogenic microbes, only a single cyclostage is highly virulent, only a single stage is toxic and only a single stage is motile, has capsules or presents a certain fermentation reaction. To these examples of correlation between a definite cyclostage and a definite physiologic reaction, apparently now one more must be added: It is only a certain cyclostage that is most sensitive to the invasive excursions of the lytic corpuscles, whether of the a or the β type.

But, if this is true, and the evidence is certainly ample, one is tempted to ask what sort of a foreign filtrable virus it is which is so handicapped in its parasitizing power that in the form of either its a or β units it can invade, and propagate in, only a single cyclostage in the development of the species? The evidence available relating to sensitivity and resistance to the bacteriophage suggests much more strongly that the assumed "invasion" of the cell is determined by a sort of physiologic reciprocity existing between the bacterium and the lytic corpuscle, a reciprocity in which there exists, on the part of the cell, in its progress through the cyclode, rather definitely timed physiologic moments of susceptibility, one for the influence of the a units, another for the β units. The biologic grounds for this contention, as well as the interpretations that proceed from it, will perhaps become clearer in the following section.

I have considered the type of cell that is sensitive to the a lytic principle, and the entirely "normal" nature of those cells that are resistant. I now come to the question of the assumed mode of generation of the lytic agent in the cells of the substratum, and will first consider only the action of the a units on the type S culture. Here the question first arising is: Is the type of cell first "attacked" the same as the type of cell in which the lytic corpuscles are generated, and from which they are eventually liberated? By the "cell type," I mean the cyclostage. This point may be expanded. Contrary to the picture presented by d'Herelle in all his writings, it is now known that the population of any pure-line bacterial culture is seldom homogeneous. Even under normal

* For example, in some recent studies with Edna Delves and Klimek on filtrable forms of the Shiga bacillus, we have observed cultures of what we have termed the "gonidial type." In early generations these are resistant to both Shiga a and β principles. After many tube-generations, however, they may return to a state in which they are again as susceptible as the S type culture from which they previously arose.

conditions of growth, the dissociative reaction is always obscurely operative, and the population is in some degree heterogeneous. It is known,
moreover, that lysis does not take place directly on the addition of the
bacteriophage to the sensitive culture, but only after several further
generations of cell division have occurred, and after the culture has
attained the critical stage. It is known, moreover, and particularly
from the reports of d'Herelle, Kuhn, Wollman, Larkum, Béguet and
others, that in the critical stage, and even somewhat earlier perhaps,
the morphologic heterogeneity of the cells is much increased; new cell
forms arise, swollen and giant forms, quite unlike the majority of the
cells of the original sensitive culture first invaded. This new form of
culture, however, is not by any means peculiar to the observed bacteriophagic reaction, as already noted by Larkum. The picture is that of the
cyclostage of the "normal" O type; and these bizarre and giant forms
can almost invariably be detected in the sensitive culture itself after
a prolonged search. Indeed, they can be increased considerably in
number by various procedures not involving in any way the addition
of prepared bacteriophage. It is, manifestly, this form of culture that is
present in abundance at the critical stage and at the beginning of the
period of liberation of the new brood of lytic corpuscles. Moreover,
there can be no doubt that to this culture type, O, are also related the
Pettenkofer bodies observed by Kuhn to be associated invariably with
the lytic reaction. Thus, to return to the original question: Can one
assume that the cells first invaded by the lytic agent are the same as
those from which it is liberated? The fact seems highly doubtful.
While (in the case being considered) the S type culture first comes
under the influence of the a principle, some of the O type cells seem
to be concerned mainly in its liberation; and, it may be added, certain
R type cells survive lysis.*

Since it seems certain that the O type cells must arise as descendants
of the original sensitive S type culture, it appears that the capacity
for liberating the bacteriophage thus becomes a function of cell multiplication, and of the attainment of a definite cyclostage. Moreover, the
changes undergone by the cells during this process of bacteriophage
generation and liberation are duplicated in almost every respect by those
modifications undergone by the sensitive culture involved in the dissociative reaction. In other words, just as the highest susceptibility to bacteriophagic invasion by the a units is the characteristic of a certain
cyclostage (the S), so the capacity for liberating the bacteriophage also

* It will be borne in mind that the R type cells also are susceptible to the influence of the
β bacteriophagic units and leave, in turn, their own characteristic resistant forms; but this
aspect of the reaction will be omitted for the present.

becomes the function of another and later cyclostage in the cyclogeny of the species (probably the O). It seems to me that these circumstances can easily imply that the thing called the bacteriophage (at least the a units) is likely to be only a specific growth stage (primary or accessory) generated by the dissociating culture at some definite point in the course of transition from the sensitive S to the resistant R culture state. It seems possible, if not probable, that this point is represented by Kuhn's Pettenkofer bodies. This conception forms the basis of the new theory of bacteriophage action which I have presented briefly in an earlier paper, and which will be considered further in the following section.*

The Artificial Production of Bacteriophage Through the Enforcement of the Dissociative Reaction.—If it is true, as I have suggested, that both transmissible autolysis and normal dissociative reactions are only different aspects of a single phenomenon, one being an acute stage and the other less acute, it might be anticipated that, just as the bacteriophage is able to incite the rapid dissociation of a sensitive culture, so also the artificially enforced dissociation of a sensitive culture should be accompanied, at some point in the process, by the liberation of the bacteriophage. In cooperation with my student, Klimek, I [170] have obtained this result. My conclusions are based on our experiments dealing with the artificial production of the bacteriophage from Shiga, coli and typhoid cultures under the inciting action of sterilized pancreatin extracts, the results of which will be presented in detail later. Using normal, sensitive cultures of the organisms mentioned. we found it easily possible to cause regularly from the Shiga culture, and less regularly from coli and typhoid, the generation of typical and highly active lytic agent merely through forcing the dissociative reaction in the following ways :

By use of pancreatin extracts, unheated.
By use of pancreatin extracts heated at 60 C. for one hour.
By use of pancreatin extracts heated at 120 C. for twenty minutes.
By use of trypsin solutions heated at 60 C. for thirty minutes.
By use of trypsin heated at 120 C. for twenty minutes.
By use of peptone solutions heated at 120 C. for twenty minutes.
By aging broth cultures for fourteen days at 8 to 10 C.

* These considerations have dealt mainly with the reaction between the a bacteriophage units (producing the large areas) and the S type culture. But it is also clear that there must be explained not only the relation between the a units and the culture which arises as a resistant to the action of the β units, but the relation between the β units and the culture that arises as a resistant to the a units. Regarding the latter two points, little can be said at present. These matters have received almost no study except from Gratia, and, indeed, the exact differences between the culture type resistant to the a principle (R) and the culture type resistant to the β principle (R²) are not known. Moreover, the relation between R² and the smooth, S type culture is not known. It is a somewhat curious circumstance that, so far as present data afford an opinion, the a principle seems to act on both the S and the R² types of culture, while the β principle acts on S and on R¹. The whole problem of the a and β units and their susceptible substrata demands much more study.

In addition we obtained similar results, though only by a longer series of filtrations (from 14 to 20), merely by growing the sensitive cultures in their own filtrates, with the occasional addition of fresh broth, as was also done for the cholera vibrio by Nobechi.

In nearly all of these tests a striking and significant phenomenon occurred. Usually, during the early series of filtrations and regrowths of the sensitive culture, a definite dissociation of the culture (as indicated by the presence of the transitional colony types) was not observed. After a certain number of filtrations, however, but before the first signs of the presence of the lytic agent could be detected in the filtrates, the cultures on the test plates began to manifest a marked change; they revealed the early and characteristic symptoms of the dissociative reaction. Klimek was the first to note this peculiarity in the Shiga cultures; and, when it appeared, was able to predict with fair accuracy (usually four times out of five) that, within the next two or three filtrations, the lytic agent would make its first appearance. If the early dissociative changes in the culture were noted, for example, in the plate test of the fifth serial filtrate, the bacteriophage was fairly sure to make its appearance in considerable abundance in the seventh or eighth filtrate. In these cases it would therefore appear that the first appearance of the lytic agent in the cultures was distinctly heralded by the preliminary signs characteristic of the dissociative reaction. Thus, although one may not yet be justified in concluding that the dissociative reaction was the cause of liberation of the lytic agent, it is at least clear that the two phenomena were closely associated in point of time and sequence.

This view draws support from the important observations of Kuhn on the significance of the Pettenkofer bodies. These were developed in abundance in cultures grown on agar containing 0.1% lithium chloride, as shown earlier by Maassen. In 1922, Kuhn succeeded in obtaining lytic filtrates from the washings of cultures grown on this medium, but not from cultures grown on common agar. Only cultures grown on lithium chloride yielded the Pettenkofer bodies; and only cultures containing these overgrown cells yielded the bacteriophage. From these observations Kuhn merely concluded that the presence of the Pettenkofer bodies "favored the production" of the lytic agent; their presence prevented the further evolution of the bacteria. Kuhn did not relate the "Pettenkoferia" or his type B, C or D substrata to the dissociative reaction or to the cyclogeny of Enderlein. His clear description of the Pettenkofer bodies, however, makes it evident that these were a phase in the development of the O type culture, in which similar forms have been described by many investigators; and their origin, in some cases at least, traced to cell conjugations. Just as Kuhn has concluded that the bacteriophage reaction, as represented by the views of d'Herelle, becomes only an aspect of the phenomena of the Pettenkofer bodies, I have reached the view that it is merely an aspect of the problem of microbic dissociation. As I understand Kuhn's course of thought, we have been converging, from different angles, on the same problem; and have arrived at somewhat similar conclusions.*

Conclusions.—Evidence for the conclusions presented in the foregoing pages may be found in numerous and scattered contributions the majority of which I have listed in an earlier paper. Many of the observed results I have been able to confirm in my own laboratory, working with dissociating cultures of many species, and with numerous strains of the lytic agent. I believe that the natural conclusion which can be drawn from these data is that the outstanding reaction of bacteria which characterizes the phenomenon of transmissible autolysis is not the lysis and disappearance of the cells of the sensitive culture, but the series of progressive transformations in culture type, in which certain stages disappear while new forms make their appearance. The acute lytic reaction, on which d'Herelle and most other workers, except Eastwood, Kuhn, Fischer, Breinl, Hoder, Kauffman and a few others have focussed their attention, may be regarded, I believe, merely as the side-issue of a phenomenon the roots of which penetrate more deeply into the complicated reproductive mechanism of the bacterial cell than any investigators have suspected. Thus does the bacteriophagic reaction become merely one aspect, but probably the most important aspect, of the greater phenomenon of microbic dissociation.

11. A THEORY OF TRANSMISSIBLE AUTOLYSIS BASED ON THE DISSOCIATIVE REACTION (HOMOGAMIC THEORY)

From what has gone before, the reader will gain the grounds for my conclusion that the bacteriophage is not a catalyzer, nor an autolysin, but a living thing, corpuscular in nature. The question now arises—if the agent is living and corpuscular, is one necessarily committed to d'Herelle's view that it must represent a foreign filtrable virus, or to the less plausible theory of Bail that it is a chromosomal "fragment" of the bacterial cell? Or to the view of Kuhn and of Koch and Ziegenspeck that it is the "spore form" of a myxomycete capable of parasitizing and destroying bacterial cells? There is still, I believe, another possibility; and one which has not heretofore been presented (although the views of Kuhn are closely related), probably because the biologic reactions on

---

* This refers to Kuhn's earlier conclusions. Later he embraced the view of a myxomycetean parasitism to explain the nature of the lytic elements.

which it is based have long been ruled out of bacteriology by the dominating monomorphic views of the nature of bacteria held by probably the majority of present day bacteriologists. This new conception involves in its major aspects the reciprocal behavior, in the reproductive mechanism of bacteria, of certain physiologically differentiated cell types occurring as definite cyclostages (primary or accessory) within the bacterial culture. But this is a view which will not find acceptance among those bacteriologists still subservient to the dictates of monomorphism; nor even ready acceptance, I suspect, among others who have not observed from personal experience the remarkable phenomena occurring in the dissociative reactions. In view of the impossibility of presenting within the limits of this review more than a brief statement regarding many of the important observations bearing on dissociative changes in cells and in colony type, I must again refer to my earlier review on microbic dissociation as furnishing the elementary point of view from which the following presentation of my theory may appear at all logical, or perhaps even understandable, to the average reader.

The Significance of Secondary Colony Formation.—When speaking of a "pure culture" of bacteria we usually picture to ourselves, as d'Herelle has done, a growth in which the individual cells bear close resemblance to one another in respect to morphologic and physiologic characteristics. As we say, the population of the culture mass is homogeneous. That such a conception, however, can no longer hold should be indicated to us by many circumstances, but particularly by the production of secondary or daughter colonies within the mass of mother culture. These pictures, already observed for many years, but at the same time possessing a curious novelty to many bacteriologists when seen for the first time, appear at some time in all bacterial species, and clearly indicate (even if other evidence were not available) that the culture is not by any means homogeneous, but heterogeneous, with respect to its cell population. Certain centers of growth, sometimes few and sometimes many, exist in such cultures in which cells of a type different from that of the mother culture as a whole, in both form and function, are pursuing their individual activities. In some species several distinct forms of secondary culture can be recognized, each differing from the others in cell form, manner of growth and physiologic capacity. If these new growth centers succeed in opposing the mother culture to a sufficient extent, they appear as knobs or as papillae on the surface of the primary colony, and so become differentiated from the surrounding mass. In other cases, they may not make any impression on the surface of the primary colony because they lie deep within it—sometimes even penetrating the substratum as observed by Faith Hadley [351] in Streptococcus fecalis. On the other hand, if they do not succeed in opposing the mother culture, they may be overwhelmed and apparently disappear in the apparently homogeneous culture background.

While it is the former reaction that usually occurs and results in the clear differentiation of the secondary culture type, which often shows greater vitality than that of the mother culture itself, this does not always happen; it depends on the culture type, the cyclostage represented by the primary and secondary cultures, respectively. In some cases the disappearance of the secondary colonies may occur; and this disappearance is sometimes rapid and complete.

A paper dealing with secondary colony formation in B. anthracis, and possessing much significance for the present considerations, was published by Preisz [288] in 1904, and the observations have several times been confirmed in later years—in the last instance by Nungester [363] of this laboratory. The observations of Preisz were as follows: In colonies of B. anthracis secondary colonies containing a different form of culture from the original were seen to arise. These colonies in time disintegrated and left round or half-moon shaped areas ("Locher") which were almost bare of growth. In these areas, after a time, still another generation of colonies (tertiary) often arose. Preisz was inclined to explain these curious reactions as due to the generation, and subsequent destruction by lysis, of a certain type of secondary colony which, in turn, gave rise to a third colony form (tertiary); only the latter survived in the now empty "Locher." Similar pictures were observed later by Pesch,[275] Katzu [209] and Brown and Basaca.[60] They have also been reported for Monilia by Sonnenschein [314] and lytic phenomena characterized by somewhat similar appearances, though lacking the tertiary colony formation, have been observed in B. pyocyaneus by Canzik,[84] Sonnenschein and many others including myself.*

It is scarcely necessary to call attention to the suggestive relation existing between the "Locher" of Preisz and the lytic areas in cultures of members of the colon-typhoid-dysentery group. In both, round or slightly irregular, bare areas are produced in the mother culture; and in both this reaction is followed by the generation of new colonies that

* In my paper on dissociation I have called attention to the possibly significant circumstance that in B. anthracis and in B. pyocyaneus the erosive or lytic phenomena here described represent the only evidence there is suggesting bacteriophagic reactions (on solid mediums) in these species. In none of the species characterized by the classic lytic phenomena of d'Herelle is an exactly similar erosive reaction observed, although I have recently come on a culture of B. typhosus that is both lytic and lysogenic. The significance of these circumstances I have considered in my earlier paper.

are apparently resistant to further disintegration or lysis. In the case of Preisz, these colonies were clearly tertiary, while in the works of d'Herelle and others, apparently analogous colonies arising in lysed Shiga cultures have been described as secondary.

Observations of this sort, supplemented by those on Monilia, and on B. pyocyaneus in which a lytic agent is undoubtedly concerned, cannot fail to suggest the possibility that the lytic areas of d'Herelle, as observed particularly in cultures of members of the intestinal group, represent the site of disappearance of localized centers of a slightly stable secondary growth; moreover, that the group of resistant colonies which arise on the lysed areas are in reality tertiary, rather than secondary, colonies.

In attempting to explain the nature of lytic areas on the grounds just presented, the question naturally arises as to the possible nature of the culture or cell type which presents the "vanishing reaction." It does not seem to be the smooth, sensitive type, nor the rough and resistant R form.* There is, however, a third culture type now recognized by many investigators, both before and after the recognition of the dissociative reaction, and this is the O or "intermediate" type, occurring in the transition from S to R. This form of culture is highly variable in its cell population and in its colony behavior, and often seems to be lacking in the S to R transformation. In this matter, however, I believe we may agree with Bernhardt that it is always present, though often unrecognized, in the culture mass. Furthermore, this cell type seems to come into prominence just preceding the critical stage and preceding the liberation of the lytic agent in cultures undergoing lysis. It has thus seemed to me that especially within this group of intermediates, in which considerable variation exists among themselves, one may well look for a certain type of cell with some specialized function, and whose reproductive behavior underlies the lytic reaction.

Following further the lines of the present hypothesis it may now be assumed that there exists in bacterial cultures a type of cell which, through some sort of transformation, can become the liberator of the bacteriophage and, thus, the determiner of lysis; also, if the present considerations may be limited to lysis on solid mediums, the determiner of the lytic area. Here one must view with much interest the Pettenkofer bodies of P. Kuhn. But it is known that the normal, sensitive

---

* It is necessary here to call attention to the fact, considered in greater detail in my earlier paper, that the common laboratory culture of B. anthracis (the Medusa head type) is not the S but the R form. In most other common species, such as B. dysenteriae, the S type culture is the common or "normal" form. This point has received further consideration by Nungester [**] of this laboratory.

cells, unaided, do not commonly, at least, acquire this new capacity. Unaided, or unmodified, they still develop, but not in the direction of lysis; they merely maintain the normal, sensitive type. But it is also known that when a dilute lytic filtrate is added to such cultures on agar lytic areas result. From this it is clear that something in the lytic filtrate changes the course of development of these sensitive cells, forcing them into a new type of culture, namely, that which at the same moment both undergoes and engenders the lytic reaction.

The Lytic Corpuscle as a Stage (Primary or Accessory) in the Cyclogeny of the Bacterial Species.—The question now arises, what might be the nature, or the mechanism of action, of an ultramicroscopic and filtrable corpuscle which, assuredly, has its first origin in the bacterial cells, and which can determine so marked a change in the developmental trend of the majority of the sensitive cells of the substratum? If it is not the corpuscle of a filtrable virus, there is only one aspect of bacterial behavior in which one might hope to find a protoplasmic unit conforming to the requirements of the bacteriophage; and that is the reproductive mechanism. This is a field of bacteriologic study in which we have too long been content to accept simple fission as the one and only means of propagation in these simple forms. But our eyes are being gradually opened to fundamentally new conceptions, namely, that the reproductive processes of bacteria are in reality complex, and that there are here involved mechanisms and forms of culture about which little is yet known. In a form associated with some of these processes, I believe, the real nature and form of the bacteriophage must be sought.

If there exists, among the cell elements supporting these dimly discerned, reproductive processes, a minute corpuscle which has the power of inciting not only cell proliferation but also a certain sort of cell "disintegration," and which, at the same time, increases at the expense of the process which it instigates, it is natural to postulate such a corpuscle as an element possessing some sort of fecundating significance. It must be able to effect a conjugation with, or perhaps fertilization of, the young, susceptible cells. In such a case, the corpuscles themselves might not have the power of independent propagation in any medium. They might possess only the ability to incite the fecundated cells, not merely to development (for that will doubtless occur in any case), but to development in a certain line; or, to employ the phrase of Hort,[352] they would instigate the "reproductive explosion." Such a phenomenon would not be new in biology, for certain instances are already known in which the future mode of development and reproduction is determined by the presence or absence of opportunity for fecundation. The cell, whether fecundated or unfecundated, will develop; but the results of the process will be different in each case.

To the majority of bacteriologists, and probably to all biologists except perhaps the mycologists, this view will doubtless appear to represent unwarranted speculation; and I freely admit that it is not yet well supported by facts relating to bacterial reproduction, although phenomena approaching this interpretation have been reported by Enderlein,[117] Almquist,[8] Kuhn, and by Koch and Ziegenspeck. It is necessary, however, for the further study of the bacteriophage, to have a working hypothesis, and the present one does not experience the disadvantage of being destitute of support. In the limited space available I cannot present this evidence further than to point out that the views of the mechanism of bacterial reproduction are at present experiencing modifications of great significance. Some of these points may be noted.

While the notion has long persisted that "bacteria multiply by simple fission," and this process only, I believe it is now sufficiently established that several other means of reproduction may intervene, at times, in the life history of the culture or species. Among these accessory means of reproduction can certainly be listed: (1) conjugation and zygospore formation, followed by endosporulation; (2) gonidia formation (macrogonidia and microgonidia); (3) a process of budding which may or may not be related to the formation of gonidia; (4) propagation through the formation of symplastic aggregates (the "bacterial plasmodium" of Almquist). In addition to these, Kuhn has described the curious reproductive behavior of the Pettenkofer bodies and Enderlein has described what he believes to be true sexual reproduction (in the cholera vibrio), involving differentiated sex elements, the oites and the spermites. Regarding the latter, judgment may well be suspended, although I do not regard the phenomenon as improbable. I mention these circumstances merely for the purpose of indicating that one or more mechanisms, designed for the attainment of amphimixis (heretofore denied to the bacteria as a group), certainly exist in bacterial reproduction. Through the work of Hort, Almquist, Jones, Löhnis, Mellon, Kuhn * and Enderlein, the fact must be regarded as established, although many of the details are still lacking.

Application of the Theory to the Facts of the Reaction.—Bringing the more important aspects of this theory, which I have for obvious reasons termed the homogamic theory (Greek, ὁμός, self, and γάμος, marriage), into harmony with the known facts of the bacteriophagic

---

* It should be added here that Kuhn, himself, does not admit that his own observations support this view.

reaction is at least not more difficult than applying the hypothesis of d'Herelle. It is scarcely feasible, at this time, to examine its bearing in detail, but a few possible questions may be anticipated. Some of these points have been considered in an earlier paper.

First, how can such a theory account for the fairly definite relation between the number of lytic areas produced and the amount or concentration of the bacteriophage suspension employed? Considering for the present only the S type culture and the a units, it is because, in a fairly homogeneous S type culture, the majority of the young cells present are susceptible to the fecundating stimulus derived from the lytic corpuscles. Only the cells receiving this stimulus will have their development so modified that they are able to give rise to the cells that later actually liberate the lytic units. The cells that do not receive the stimulus may still develop, but along another course.

Why should the lytic reaction occur only when the bacteriophagic corpuscles are added to young cultures? It may be that unless the fecundation process occurs when the cells are young it cannot occur at all. If delayed, the course of development is already set in another direction. In many of the cases in which parthenogenesis occurs there is a time limit beyond which fertilization cannot be effected, thus leaving unmodified the primary course of development.

Why does lysis not occur in bacteria suspended in salt solution? It is because the cells must multiply before they can attain that cyclostage which is capable of liberating the new brood of corpuscles; and the cells cannot multiply appreciably in salt solution or in nonnutritive mediums. Passage of the cells through at least a part of the normal cyclode must occur before they can attain the threshold of lysis.

Why is an R type culture (arising from normal dissociation) more resistant than the S type culture to the a lytic action as shown by the reduced size and number of the lytic areas, the greater number of secondaries and the failure of complete clearing in liquid mediums? This is due to the fact that the R type culture has entered a state in which the number of susceptible cells (for the a units) is much reduced, and the cyclostage represented by the predominant type (R) is not susceptible in the same degree to the same fecundating stimulus, although it is known to be susceptible to the β units. Multiplication of the lytic corpuscles, in other words, depends on the union of certain cells existing in a definite, though transitory, cyclostage.

Why, if the lytic corpuscles are present (as should be expected) in small numbers in normal, sensitive culture, is the lytic phenomenon not manifested more commonly in such cultures? To this I would say that the reaction undoubtedly occurs frequently in such cultures; and that, by appropriate methods, it may sometimes be raised to the level of the typical bacteriophagic reaction. This may be responsible for "spontaneous" autolysis and for the sudden loss of stock cultures not infrequently experienced by bacteriologists.

How does this theory account for the development of the secondary, resistant culture and the mucoid transitionals? It does so by eliminating the cells of the sensitive type, thus allowing to come into the foreground those cell forms which, at the time of introducing the lytic agent, were already embarked on a different developmental route in the cyclogeny of the species. It also seems probable that some of the R type cells are generated in some still unrecognized manner in the actual lytic reaction.

How does this theory explain the apparent adaptation of the bacteriophage to unfavorable conditions of environment, as, for instance, an acid medium? It is permissible to believe that whatever degree of adaptation the bacteriophagic corpuscles experience is transmitted to them from the mother cells, which have already survived the influence of the new environment. The process is, therefore, not adaptation, but selection. When the corpuscles generated from the selected mother cells are called on to reinstigate the lytic reaction, they reintroduce into the germ plasm of the species the qualities of greater resistance which they have inherited and can, accordingly, transmit. Thus, the actual "adaptation" is not accomplished by the lytic corpuscles themselves, but by the germ plasm of the mother cells. It may be recalled that we do not speak of the adaptation of sperm cells, although we know that they can perpetuate the genetic qualities determined by the so-called adaptation of the somatoplasm.

Finally, how can this theory explain the fact that a certain strain of bacteriophage, on first isolation and without "adaptation," is able to influence, not only the homologous cultural substratum, but also cultures of related species? The Shiga bacteriophage, for example, may be able to produce plaques, not only in Shiga culture, but also in substrata composed of B. coli or B. typhosus. To answer this difficult question involves an even more far-reaching speculation, but I may put the matter in the following form. Is one at present justified in denying the existence of bacterial hybridization? Almquist [2] believes that he has adduced evidence which supports this view, although his conceptions and conclusions have been accepted by few bacteriologists. With what is known of hybridization among higher plant forms, however, one may not safely regard the phenomenon as a biologic impossibility among bacteria; and, if it could be accepted, an answer to many instances of curious behavior of bacteria, and especially in microbic associations, would be obtained. Although I can see little ground for accepting such a view at present, only on such a basis can the heterologous action of the lytic corpuscles be explained in the terms of the present theory.

If there are some circumstances which the homogamic theory explains as readily as does d'Herelle's conception, or that of Bordet, there are also some points which it does not explain. The following may be noted, and other inadequacies will undoubtedly soon be pointed out.

Why is it that a bacterial culture resistant to one strain or form of bacteriophage is susceptible to another strain?

How can one account for the great viability of the bacteriophage corpuscles in sealed tubes?

How can one explain the facts relating to the "large area" (a) and "small area" (β) principles, which appear to be so individualistic and reciprocal in their reactions?

These questions are difficult to answer, but I do not see that they are answered any more readily by resort to d'Herelle's conception; and they certainly cannot be answered on the basis of Bordet's hypothesis

In concluding this aspect of the problem, I wish to mention one or two matters which, although they do not serve as evidence for the theory just considered, are suggestive in demonstrating that there exist, in the dissociation series, certain forms of culture of common bacterial species about which little is known, but which may play a part in the bacteriophagic reaction. Bacteriologists often fail to secure growth of a culture on a certain medium on which the organisms have previously grown and conclude that the culture is dead. Species are known, some of whose cyclostages grow slightly on agar but not visibly in broth. Bernhardt,[353] for example, studied a colony of B. typhosus which manifested this characteristic. Seeding from agar to agar yielded growth, but seeding from broth to agar did not. Soule[352] has also described for B. subtilis a form of colony ("phantom type") which would grow on agar with great difficulty, but not at all in broth. Merely suspending the culture in broth for a few moments for dilution purposes seemed to destroy the culture, for it could not thereafter be recovered. Dilutions made in salt solution were, however, viable and, when seeded back to agar, would grow. We have recently produced a filtrable form of the Shiga bacillus. At times this culture presents no visible growth on agar or in broth, although it seems to propagate itself under these conditions. At other times, it presents in broth either a slight mucilagenous precipitate or faint opalescence, but yields no visible growth on agar.

These and other results obtained by us lead me to the opinion that failure of visible growth in broth or on agar serves as no criterion of the absence of living and propagating bacteria.

Another observation of interest was made by Enderlein. He noted, as have many others, that by seeding from old agar cultures to fresh agar, in many cases growth did not occur. Under such circumstances, one usually concludes that the culture is dead. When, however, he transferred some of the old culture to fresh broth, let the tube stand in a lighted window for a few hours and then seeded to agar, growth appeared. According to Enderlein, this phenomenon concerns a sexual phase of bacterial reproduction, the details of which may be passed over. All these observations should, however, make us cautious in concluding that bacterial forms, in some cyclostage, are absent from a medium or a tissue merely because growth fails on the common culture mediums. To make more specific application of these facts, one may by no means conclude that, because often signs of bacterial growth in or from bacteriophage suspensions cannot be observed, living bacterial forms of some sort are absent. Our notions of the bacteria have naturally developed about the forms that are easily observed by common microscopic methods, but many investigations within recent years enforce the conclusion that invisible stages are by no means uncommon, and that they may play important rôles in a variety of obscure reactions, not in the culture tube alone, but also in the living body.

Conclusions.—The apparent failure of the older theories to account for many of the observed facts in the bacteriophagic reaction, and the numerous indications that the phenomenon is closely bound up with the dissociative reactions of bacteria, have thus led me to attempt to formulate a working hypothesis resting on a basis different from those supporting either the conception of d'Herelle (virus theory) or that of Bordet (nutritive vitiation theory). This attempt has resulted in the formulation of the so-called homogamic theory in which it is postulated that the bacteriophagic corpuscles represent either stages in the cyclogeny of the species of the substratum (or of a closely related species, containing at least some of the common O or R antigens), or units of accessory physiologic significance, such as fecundating elements of some sort. This theory is based on known facts relating to the dissociative reaction; and, it must be added, unfortunately, on some views for which at present proof is not at hand. Although this theory has the advantage, I believe, of furnishing a more satisfactory explanation of many aspects of the reaction than can be obtained from other hypotheses, and certainly does not involve a greater degree of speculation, I can regard it for the present merely as a new working hypothesis which may serve the purpose of suggesting a new avenue of approach to the bacteriophage problem. Whether the operation of the mechanism which I have proposed is eventually demonstrated as true or false, I am, however, confident of one essential fact—that, when the actual mechanism is discovered, it will be found to lie, not in the field of free, disruptive autolysins, nor within the territory of Myxomycetean parasitism, nor yet in the category of the foreign filtrable viruses, but within the realm of dissociative bacterial behavior. That much I regard as certain.

12. THE BIOLOGIC SIGNIFICANCE OF TRANSMISSIBLE AUTOLYSIS

The assumed biologic significance of the reaction is naturally interpreted by different workers in terms of their conception of the causal factor. For d'Herelle it represents the act and consequences of parasitism, and, therefore, does not possess an unusual, intrinsic, biologic significance. For Bordet it represents a disturbance in the metabolic equilibrium; and, since the reaction may be instigated either within the cell (active reaction) or from without the cell (passive reaction), the phenomenon possesses intrinsic as well as extrinsic significance. Both these workers have presented views dealing with the rôle played by the lytic agent in the production of bacterial mutations and (Bordet) in controlling the destiny of the species. Still other investigators, as Breinl, Fischer and Hoder, also Eastwood and Kauffman, have grasped the view that, in respect to the variants or "mutants" produced, the bacteriophage seems in some way to be related to "normal" variation and mutation phenomena in bacteria; but they have not detected the relation of these phenomena to the orderly manifestations of the dissociative reaction. Since the views of these workers have received consideration earlier, they will not be referred to again in the present section. In this group there remains only the conception of the Wollmans, who have reinjected into the bacteriophage problem certain points dealing with the "hereditary" features in a manner quite different from that of Bordet.

REFERENCES

Twort [122] (considered many possibilities, but embraced none).

D'Herelle [203] (bacteriophage as the agent responsible for all bacterial mutations).

Bordet[40] (bacteriophage autolysin as a substance whose abnormal fluctuations determine mutations, but whose normal activity controls the destiny of the species).

Wollman and Wollman[242, 243] (possible relation of bacteriophage to their phenomenon of "paraheredity"). Also Wollman.[335]

Wollman[330] (culture modification or lysis through "hereditary factors"—paraheredity).

DISCUSSION

From his primary experiments on the autolytic reaction in staphylococcus Twort could not come to definite conclusions regarding either the cause of the phenomenon or its significance. Among various possibilities he considered that the manifestations might deal with a new form of living stuff. In this connection, Wollman,[335] also, though leaning more strongly toward views related to his "paraheredity," earlier expressed the notion that it may not be possible to find the nature or significance of transmissible autolysis within the confines of biologic facts already known.

More recently, Wollman[336] has made his "paraheredity" the basis of his newer hypothesis of the "hereditary 'factors.'" The conception here involved concerns itself with a general biologic problem of large dimensions, although one that has not aroused much interest in recent years; and his conclusions are, perhaps, no less radical than those involved in my own homogamic theory of bacteriophage action. As has already been seen, the significance of his hypothesis lies in his attempt to establish a physical basis for the hereditary transmission of characters through the external medium, and thus to submit a mechanism for the inheritance of acquired characters among bacterial forms.

Aside, however, from its relation to the dissociative reactions of bacteria, which has already been sufficiently considered, the rôle of the bacteriophage in the production of assumed mutations is probably of greatest interest.

Apart from regarding the phenomenon as an instance of parasitism of bacteria by a foreign virus, d'Herelle conceives the principle as the agent responsible for all bacterial mutations. He states:[193e] "It is indeed probable, as various investigators have suggested, that all of the fixed mutations occurring in bacterial species are produced through the action of the bacteriophage."

It must also be noted that Bordet likewise discovered in the bacteriophage a chemical agent elaborated by the cells and possessing power of directing the trend of bacterial evolution. In this manner "it controls the destiny of the species." For him, it thus possesses a regulating influence, which, when the agent functions normally, keeps the culture true to type. He even can picture an analogous substance working in the cells and tissues of higher animals.

To these views attributing to the bacteriophage the power of producing bacterial "mutations," or (if functioning normally) of keeping the culture true to type, much the same answer must be given. It cannot be demonstrated experimentally that the bacteriophage, as understood by either d'Herelle or Bordet, is able to produce, in bacterial cultures, any more variable or striking modifications than those already recognized as occurring spontaneously (that is, without recognized bacteriophagic intervention) in many bacterial cultures when placed under certain growth conditions, such as may be determined by chemical and physical agents of great variety. "The important point—indeed, the crucial point—which d'Herelle fails to comprehend is that the cyclogeny of a single bacterial species embraces many strictly 'normal' forms of culture growth, each one of which is endowed with different biochemical, serologic and antigenic characters; and, it may be added, with quite different degrees of resistance to the action of the bacteriophage."[168f] In all these cyclogenic changes the bacteriophage (as d'Herelle understands it) plays no other rôle than to accelerate such reactions as the culture is ready to accomplish by itself in a less energetic manner.

This criticism cannot be applied in equal degree to the views of Bordet, for his conception possesses the great merit of having focussed on a point which has, from the beginning of his studies, managed to escape d'Herelle; namely, the circumstance that the bacteriophagic reaction is only a normal reaction driven to the extreme. But the work of Bordet has another aspect of significance which is lacking in the considerations of d'Herelle: Bordet has recognized, and made use of, experimentally, some of the heterologous culture types which occur in the strictly "normal" culture; he has thus concerned himself with some of the facts and phenomena of the dissociative reaction, although he has not surrounded them with this interpretation. In any case, however, he has penetrated much more closely to the biologic basis of the bacteriophagic reaction than d'Herelle has been able to do, preoccupied with details of the virus theory. Fortunately for the great mass of experimental data assembled by Bordet to support his views, the failure to discover in his variants the perfect analogy with quite normal stages in the cyclogeny of the species (rather than abnormal forms that required "regulation," as he regarded them) does not endanger the great value of his studies. They will, in time, I believe, receive a new interpretation free from the implications of the bacterial autolysins, or of any other purely chemical product of the bacterial cells.

Thus, to summarize regarding the relation of the bacteriophage to
bacterial mutations, I believe that there exists at present no unequivocal
case of mutation recognized among the bacteria, whether produced by
the bacteriophage or in any other manner. And this is due to the circumstance that a knowledge of the limits of cyclogenic variation within the
species has not yet been acquired. Not until this has been done can one
recognize the permanent, hereditary departure from the specific cycle.
The alleged mutation producing rôle attributed to the bacteriophage by
d'Herelle, when fully analyzed, is found to be one which merely hastens
the dissociation of the sensitive culture into its component cyclostages,
some of which, it is true, are remarkably stable. Since, however, a
similar dissociative reaction can occur in sensitive cultures to which a
foreign principle has not been added, one cannot conclude that the
bacteriophage (at least in the d'Herelle sense) is necessary even for such
an analysis. These observations lead to the further suggestion, that the
dissociation of a susceptible culture by the bacteriophage, and the generation of the bacteriophage by the dissociating culture, are concurrent
and reciprocal phenomena, united in some still unrecognized manner in
the reproductive mechanism of the species.

*Transmissible Autolysis as the Pathologic State of a Normal
Reaction.*—If the foregoing fact is true, and it is supported by a certain
sort of evidence, what can one conclude regarding the interpretation of
the bacteriophage reaction as a pathologic process? If we refer to the
laboratory reaction, the classic reaction of d'Herelle, it would appear that
the phrase of Bordet and Ciuca characterizes it perfectly. It is "a normal reaction carried to a pathological extreme." It is the segregation
and intensification, by laboratory methods, of a form of reproductive
behavior already inherent in all bacterial species. The important question that confronts one is, therefore: What is the actual nature of this
mechanism, and how does its "intensification" arise?

Considering the phenomenon from this point of view naturally
requires the recognition of a much wider range of manifestation of
bacteriophagic reaction than has been permitted by d'Herelle, who draws
about the phenomenon such limitations as those circumscribing the
Shiga reaction. For this reason he is not able to accept the phenomenon
of Twort involving the staphylococcus, of Pesch, Katzu or Pico for
B. anthracis, of Canzik and others for B. pyocyaneus or, probably, of
Sonnenschein for Monilia. These reactions are all regarded as examples
of "bacterioclysis," another "disease of bacteria." Although further
information is not available regarding the possible nature of this disease.

"bacterioclysis" serves as a convenient repository for all such troublesome phenomena. Some investigators, as Bordet and Gratia, have definitely accepted the Twort phenomenon as analogous to d'Herelle's reaction, while others impose distinctions. Bordet's acceptance is based on the following facts: that the Twort reaction affects only living bacteria; that it is indefinitely transmissible, and that the agent is recoverable in filtrates.

D'Herelle's conclusions regarding the lytic reactions of B. pyocyaneus are the same as those for Twort's staphylococcus lysis. In this connection, one of d'Herelle's arguments for the view that pyocyaneus lysis does not represent the true bacteriophage reaction is of special interest. It is because this reaction in pyocyaneus is so often spontaneous. But if, during the past few years, any one fact of importance has been revealed regarding the origin of the bacteriophage, it is that this principle develops spontaneously in bacterial cultures if they are surrounded with a favorable environment or receive an adequate stimulus. It is scarcely to be assumed that the environmental conditions which are fitted to reveal the bacteriophage in one species will do it in all; and it is reasonable to believe that some species, characterized by a different cyclogeny, should reveal the reaction more readily than others; moreover, that they should reveal it in a somewhat different manner. As examples of this we may regard the lytic reactions of B. pyocyaneus and B. anthracis, and probably of still other species for which, as yet, no bacteriophage has been discovered. From my own experiments on the lysis of B. pyocyaneus I have concluded that this is a species in which the degree of transmissibility of the principle is low; the bacteriophage is highly spontaneous but rather weakly transmissible. It may eventually be shown that there is a connection between these features, and that a high degree of transmissibility will be found only among those species in which the spontaneously lytic reaction is less commonly observed. Since there has been some controversy over the position of pyocyaneus lysis, I may restate my own conclusions as follows: In this species, lysis is both spontaneous and transmissible; it is expressed by both inhibition and lysis in broth and on agar; the lysis on agar is characterized by areas of erosion; the disappearance of the sensitive culture is followed by the generation of resistant forms which differ from the original morphologically, biochemically, in colonial features and in virulence. These characteristics, it seems to me, are sufficient to demonstrate that this phenomenon is an example of the bacteriophage reaction.

What has been said of B. pyocyaneus may also hold true for B anthracis, although here further study is required. Although Pico,[281] in 1922, reported a lytic agent for this species, d'Herelle [193] does not accept this. Pesch, Katzu, Sonnenschein and Brown and Basaca have all noted the erosive features in anthrax cultures; but none of them has been able to demonstrate transmissibility of the agent, and they have concluded that the bacteriophage was not concerned. The same was true in Sonnenschein's [314] instance of lytic areas in a culture of Monilia from the throat.

There are other instances in which we find degrees of plaque formation less clearly marked than in members of the intestinal group, which seem to have been accepted as the standard with which all other pictures must be compared. In Enterococcus and in Friedländer's bacillus, other variations exist, as I have observed in many experiments on these cultures and their homologous lytic filtrates.

It seems to me that these and similar facts should make it clear to us that the mode of expression of the bacteriophage, at least so far as lytic areas are concerned, may be highly variable in degree of transmissibility, in intensity and in spontaneity. Manifestly, such overlapping grades of the reaction exist that it is impossible to draw an arbitrary line between the weaker and the stronger; or to conclude what, on the one hand, is transmissible autolysis, and, on the other, a different sort of reaction—"bacterioclysis." It seems much more in accord with the observed facts to assume that the phenomenon of transmissible autolysis among bacteria at large covers a wide range of reactions; and that the detailed character of the reaction in a given case will depend on the bacterial species under consideration; moreover, that the observed differences will vary with the nature of the specific cyclogeny.

The Possible Significance of the "Lytic" Reactions of Bacteria.— In connection with the bacteriophagic reaction determining the so-called lysis of bacteria, it seems desirable to consider briefly what may actually be involved in certain other reactions of bacteria, often termed "bacteriolysis," etc. The question that arises is whether we are justified in assuming that, because bacteria are seen to granulate and sometimes disappear from view when acted on, for example, by immune serums, these bacteria have been destroyed. Is the modified bacterial cell destroyed by the bacteriophage, or is it merely transformed into another and invisible state through the liberation of its progeny? Hauduroy and d'Herelle and Hauduroy have clearly shown the possibility of producing filtrable forms of common bacteria under the influence of the bacteriophage. Fejgin [123] has stated that "the virus of typhoid is merely the bacillus of Eberth lysed by the bacteriophage of d'Herelle." In these cases at least, there is not, therefore, an actual destruction of the bacteria, but a transformation into another state or stage of existence.

Years ago, Radziewsky [292] pointed out that in the lysis of the cholera vibrio and typhoid bacillus under the influence of normal human serum the granulation and apparent rupture of the cell did not prove that its destruction had occurred; for further culture tests showed that after several days, perhaps, a growth might appear, apparently arising from the granular débris. Similar facts were presented by Eisenberg [852] in 1903. In the older literature, also, much evidence indicates that in the Pfeiffer reaction, bacteriolysis and the bactericidal reaction do not go hand in hand. This may well cause one to question whether many instances of this sort may not represent transformation, rather than actual destruction, of the bacterial cells. The functional germ plasm may continue to exist in another form. In this respect, it may be added, the biologic conception lying behind Bail's "chromosomal" theory of bacteriophage action is not without an element of interest, although the supplementary details with which he has clothed this element of truth are of such nature that the entire hypothesis becomes incongruous and unconvincing. The fact remains, however, that Bail's observations on the bacteriophage reaction were of such a sort as to convince him that the germ plasm of the bacterial cell and of the lytic units maintained an unbroken continuity. From all these observations one may well conclude that the disappearance of the familiar forms of cultures does not always, or necessarily, imply their death; and that, before one can affirm that one knows the full biology of any species, new methods must be devised for tracing its path into the realm of what Nicolle [261] has termed the infrabacteria.

The Filtrable Forms of Bacteria.—Closely related to the question of the biological significance of the bacteriophage and of the lytic reaction is the problem of the existence of filtrable forms of bacteria—a subject which has assumed new interest in recent years, particularly as a result of the studies of the French school of bacteriologists dealing with the bacillus of tuberculosis, and following the discovery of the filtrable forms of this organism by Fontès in 1910. In relation to the bacteriophage this subject is of special significance because, upon its final solution, will depend the truth or falsity of the homogamic theory which I have proposed to explain the nature of the bacteriophage and its mechanism of action. While numerous bacteriologists readily accept the view that certain bacterial cells are able to pass Chamberland or

Berkefeld filters, the majority of present-day workers are opposed to the view that there exist, in the developmental history of bacteria, special forms whose unusually small size permits of easy and frequent filtration. The majority of those who oppose this conception, it may be observed, are those who have slight knowledge of, or belief in, the phenomena of microbic dissociation, or the significance which I [168] have ascribed to these reactions. Most of them are still laboring under the spell of an ancient monomorphism.

At the present moment it is not my purpose to review the numerous investigations which bear on one side or the other of this controversy. I wish merely to point out one or two factors which are commonly omitted in discussions and which must be taken into consideration in any conclusions as to whether filtrable forms of bacteria in reality exist. And, when I here use the term, "filtrable forms of bacteria," I do not refer to small individuals of the "normal" type of cell which are unquestionably able, at times, either because of their small size or shape, or because of some minute defect in the filter or its manner of control, to find a passage into the filtrate. I refer particularly to those specialized cell forms different from the "normal" type, not only in size but also in several other features, including biochemical reaction, serological reaction and virulence—forms, the exact nature of which (at least outside of the microgonidia) we are still far from knowing, but which may tentatively be regarded as occupying a place in the complete cyclogeny of the species concerned.

When a culture that can be obtained from a filtrate (such as a Berkefeld N or a Chamberland L^2 or L^3 filtrate) is found, on the immediate growth, to resemble the original culture morphologically and biochemically as well as serologically and in virulence, there is certainly nothing noteworthy in the circumstance; some of the smaller units of the culture probably pass through the filter and at once take up their growth, soon coming to resemble the original culture type. This no doubt frequently happens, especially with such small forms as B. prodigiosus, the Pasteurella, and perhaps with some streptococcal and spirochaetal forms. Such results naturally cannot be taken as evidence supporting the existence of filtrable forms of bacteria as this term is now employed. So far as I am aware this phenomenon has been found to occur only with cultures comprising the smallest or most slender organisms and not with relatively large-cell types such as B. subtilis, B. anthracis * and B. diphtheriae.

* In this species it may be noted that Haag (Archiv f. Hyg., 1927, 98, 271) has recently reported a filtrable stage, but that it is quite different from the typical anthrax bacillus.

The feature which altogether determines the significance of the new growth that may occur in culture filtrates is the form that the growth takes and the morphological, biochemical and serological characteristics of its visible units.   And to this one might add any change in its pathogenic power.   It is only in cases in which there plainly exist such striking differences between the old culture type and the new that we are justified in concluding that a "filtrable form of the organism" is concerned.   In other words, the forms of the bacteria that are filtrable can be recognized as possessing, for a limited time at least, characteristics far different from those of the mother culture; characteristics that are sometimes so different that it may almost seem unjustifiable to assume that the new culture is, in reality, a linear descendant of the old.   These new characteristics may be retained for a considerable time; or the new culture type may, within a few transfers, "revert" to the old form.

The time within which such secondary cultures arise in filtrates is of much significance.   In the available literature, as also in my own experience, the following circumstance may be noted.   If the secondary growth arises within 24 to 48 hours after filtration it usually resembles in all important respects the original culture.   Under such conditions there may be no reason to doubt that the appearance of these organisms in the filtrate is due either to their small size or to a defect in the filter. If, on the other hand, the secondary growth is delayed for a period of from three to thirty days after filtration, as is very often the case, the new culture type is usually found to show marked differences from the original culture, not only in the morphology of the cells but also in other important characters, sometimes including virulence.   Under these circumstances it seems logical to assume that the "filtrable form" of the organism is one that, at the time of filtration, was quite different from the original.

In many of the growths which occur after a considerable time in bacterial filtrates it is apparent that the new morphological type of cell is a minute coccus, varying in size from perhaps one micron down to the limit of vision, forms measuring $0.2\ \mu$ being common.   Such coccus or coccoid forms have been described for many species of bacteria; and I have examined them in broth cultures of B. diphtheriae, B. malleus and several members of the colon-typhoid-dysentery group. In the Shiga dysentery bacillus this form often occurs as a minute, fine-chained streptococcus.   Such coccus forms have been noted by Hauduroy and by d'Herelle and Hauduroy in various species when acted on by the homologous bacteriophage; and these writers have regarded as especially noteworthy the circumstance that, whatever the original morphological type (spirillum, rod or coccus), the ultimate form attained was that of a minute coccus, often cultivable on common mediums with the greatest difficulty and after a considerable time, if at all.

Ample evidence has been supplied during the past twenty years to suggest strongly that these minute bodies represent, in part at least, the gonidial forms of the culture. With my students, Delves and Klimek, I have produced similar forms in Shiga dysentery cultures, and we have also observed them in B. typhosus. These cultures are readily filtrable through such candles as the Berkefeld N and Chamberland L³. For a time at least they may not be cultivable on common solid mediums. In infusion broth they may or may not be visibly propagated. At the beginning they are likely to reveal, if anything, a very delicate and often viscous sediment in which may be observed a fine mycelial structure containing minute granules, such as those described by Ramsin, Hauduroy and others. Often subcultures may be carried through a long series of broth tubes none of which reveals anything that would ordinarily be construed as representing visible growth. Eventually, however, they become transformed into bodies (resembling the "regenerative bodies" of Löhnis) which yield a delicate growth on agar, giving colonies that, at the beginning, are seldom more than 0.1 mm. in diameter and characterized either by a whitish opacity or by a bluish translucency. With continued transfer these colonies become larger until, eventually (sometimes only after weeks), they attain the "normal" colony form; and with it the "normal" cytological type. As I have already shown with Edna Delves, these minute colonies (in this case the blue and translucent type) may appear suddenly and in great numbers in cultures of the Shiga bacillus directly with the disappearance of the S type cells, which a few hours before had represented the preponderant culture type in broth. Under these conditions the gonidial colonies may be of at least two types, one containing minute cocci, the other minute rod-shaped elements. The former type of colony is often rough with a serrated margin, while the second is round and smooth with an even margin. Although it is possible that these two colony forms may represent different stages of the same type of gonidial colony, this at present seems improbable, since both forms tend to remain constant in numerous passages on agar plates alternated with broth cultivations. From these and many other related observations which cannot be reported here, I am inclined to the view that the filtrable forms of bacteria most commonly encountered represent either the gonidial bodies or the regenerative elements which spring from them.*

If one takes cognizance of the facts of the dissociative reaction it might be expected, however, that the new form of culture would not remain indefinitely in this state; but that it would, through some form of further development, eventually revert to the original type. And this result has been shown to be the case in several instances—notably in the filtration tests of the tubercle bacillus, of various intestinal organisms (Hauduroy [176, 177]), of B. typhosus (Friedberger [352] and Bronislawa Fejgin [123, 125]), of B. proteus (Fejgin [126]) and of the fusiform bacillus (Mellon [256a]). And one may add here the very interesting studies that seem to demonstrate an alternation of generation between filtrable and microscopically visible stages in the development of B. tumefaciens and the virus of crown gall.

Some recent writers who have hesitated in accepting the view that there may exist a filtrable form of such pathogenic organisms as the so-called scarlet fever streptococcus have attempted to effect an explanation of the facts observed in this disease by suggesting that a distinct biologic entity in the form of a filtrable virus responsible for scarlet fever may live in intimate contact with the scarlet fever streptococcus; and may indeed at times be separated from it. The same view might be applied with equal justification to the splendid work of Alice Evans [352] on her strains of greening streptococci from encephalitis. But, in view of our present knowledge of the filtrable forms of many common bacteria, such notions merely lead us into speculations much more unnecessary and unwarranted than the thesis that I am inclined to support—namely, that the forms of ultravirus undoubtedly associated with both scarlet fever and encephalitis are only filtrable stages in the cyclogeny of the microscopically visible streptococci which have long been held as

---

* One argument against the existence of such filtrable forms of bacteria has been that, when (as in B. tuberculosis or streptococcus) laboratory animals are employed to demonstrate the presence of the virus, there is no guarantee that the germs concerned were not already lying latent ("microbes de sortie") in the tissues. This possibility has recently been mentioned by Palante and Koudriavtzeva (Compt. rend. Soc. biol., 1927, 96, p. 1218) for streptococcus infection in mice. The point has also been raised by Et. Burnet (Ibid., 1928, 98, p. 440) for streptococcus infections in rabbits. Moreover, some investigators believe that they have reason to doubt that the alleged tubercular infections in guinea-pigs, following the injection of filtrates, are due to filtrable forms of the tubercle bacillus rather than to "microbes de sortie." This is a point of importance that must be taken into consideration in all animal tests. On the other hand, it may be pointed out that animal tests are not always required for the demonstration of the existence of filtrable forms of bacteria; for continuous cultivation methods are often sufficient for this purpose. This is true for streptococcus, staphylococcus and for several intestinal species, including the cholera vibrio. It may be anticipated that the doubts surrounding the situation with respect to the tubercle bacillus will be dispelled when the original acidfast type has been recovered by continuous cultivation from the filtrable bodies alleged to be present in filtrates. This will without doubt be reported in the near future. In the meantime, the discovery of "microbes de sortie" may be regarded as militating in small measure against the conclusions derived from purely cultural studies demonstrating the existence of filtrable forms of many bacterial species.

causally related to the diseases in question. Indeed, taking a more general view of the situation, in the light of our present knowledge it becomes increasingly difficult to maintain the viewpoint that the large group of filtrable viruses now known to us represents a new and special type of organic being. I believe the time is drawing near, mono-morphists to the contrary, when we shall come to regard at least some of these viruses as representing merely virulent, usually ultramicroscopic and filtrable stages in the cyclogeny of certain bacterial species, with which we have perhaps long maintained an unsuspected intimacy. More-over, I believe that, until we cease attempting to study and to classify the filtrable viruses as isolated and specific entities, and begin to examine their possible relation to the visible culture forms that may precede, accompany or follow them, we shall make little headway in the eluci-dation of this intricate problem.

The thought may occur to some who object to such apparently radical views as I have expressed in this and other papers, that it is a highly remarkable, and perhaps impossible, circumstance that, if the ultravirus is merely a stage in the development of a bacterial culture, the virus can be perpetuated over such long periods of time without appearing to manifest a tendency to revert to the original type. In reality I can see nothing at all remarkable about this situation, for it is now well known that other distinct stages in the full cyclogeny of a culture may become stabilized for weeks, or even for years, provided the environment is favorable. Some of these cyclostages have, indeed, been regarded as "irreversible." For example may be cited the sta-bilization of either the S or the R type cultures, each bearing no resemblance to the other, but remaining without perceptible alteration over long periods of time. Properly treated, the O type culture may also retain its characteristics with considerable tenacity, although this form is the least stable of the cyclostages that have been studied. We have recently discovered a new form of culture of the Shiga dysentery bacillus which may not give visible growth on plain agar or even on blood agar, but which, in broth, can be perpetuated for weeks in the form of an invisible, or sometimes barely visible growth. These cul-tures are always filtrable and I am inclined to the view that they represent the stabilization of the microgonidial stage in the cyclogeny of the species. They appear to partake of the nature of a cultivable but invisible virus. Eventually they may revert to the original culture form. The same may be said of the "regenerative elements" which seem to spring from the gonidial forms. These, whatever their real nature, may be perpetuated on agar for many generations with slight perceptible change either in colony form or in the morphology of the individual elements.

If the culture types just mentioned are able to guard so jealously, and for so long a time, their characteristic attributes (often being forced into line again only by some highly specialized treatment, such as repeated growth in homologous immune serum), there is small need to be astonished that a filtrable cyclostage, if endowed with pathogenic power, could maintain its integrity through numerous animal passages, and perhaps through a complete epidemic of a communicable virus disease.

The influence of the bacteriophage in the production of filtrable forms.—If now we may tentatively assume that there exist such filtrable forms of bacteria which differ from the original type in many important respects, which are not observable in all cultures, but only under certain conditions of growth, and whose discovery is dependent upon their having entered the favorable cyclostage at the exact time that the filtration is performed, then what can be said regarding the relation of the bacteriophage to these filtrable bodies and to the mechanism concerned in their generation?

Although common bacterial species, existing in an apparently "normal" culture state, may sometimes reveal filtrable forms, it is a matter of common experience that they usually fail to do so; and from this we are likely to conclude that filtrable forms do not exist for the bacterial species in question. But it has been revealed by numerous investigations, among which may be noted particularly those of Hauduroy,[176, 177] d'Herelle and Hauduroy,[196] and Bronislawa Fejgin,[123, 126] that cultures placed under a weak bacteriophagic stimulus are especially likely to generate filtrable bodies; and that these filtrable forms differ markedly from the original cultures. Some, as in Fejgin's work with the ultravirus of typhoid, may be maintained in the filtrable state through many animal passages. Moreover, as a corollary to this, it may be added that, when bacterial cultures are forced to generate their own bacteriophage, as through aging or under the influence of sterile pancreatin or trypsin solutions (Hadley and Klimek [170]), they automatically enter a state in which the generation of filtrable forms is common; also that these filtrable bodies differ markedly from the original culture type. In other words, we may regard the bacteriophage as an agent under whose influence apparently "normal" cultures may be made to generate special cell types that pass readily through the common porcelain or earth filters.

The question therefore arises, How shall we regard the mechanics of this process? Does the bacteriophage cause a sort of fragmentation of the sensitive cells into minute particles which, if their charge be favorable, can pass the filter and then regenerate into the microscopically visible form? Does the bacteriophage fractionate the cells into elementary chromosomal bodies, as in Bail's conception, or into hypothetical pangen-like particles as suggested by Wollman's view? All these conceptions are very improbable. In earlier sections I have attempted to point out that there exists no available evidence that the lytic agent inflicts such violent physical injury on the bacterial cells that the stream of germ-plasm is annihilated. The real and foremost function of the bacteriophage is, rather, to determine a transformation in the nature of the cellular elements, causing the apparent annihilation of some forms that may previously have constituted the larger part of the bacterial population and bringing into evidence others which were previously non-existant, or present only in small measure.

To conclude, therefore, if the chief function of the bacteriophagic stimulus is to produce cell and culture transformations; and if, during this process, we find that the culture concerned possesses a larger number of those bodies which readily pass the common filters, it seems reasonable to conclude that the generation of filtrable forms of bacteria is the product of a certain sort of culture transition from one cyclostage to another; and that the chief agents—perhaps the only agents—capable of accomplishing this end are the lytic corpuscles themselves. Whether cultures can enter a state in which some of the cellular elements become filtrable, but unaccompanied by the presence of the bacteriophage or some physiological counterpart (artificially added or spontaneously generated), we do not yet know. At present, all we can state with assurance is that, when filtrable culture forms arise, they seem to be most commonly associated with bacteriophagic influence and with dissociative reactions.

Conclusions.—In concluding these references to the biologic significance of the bacteriophagic reaction it becomes apparent that one is confronted with three possible views. The first (that of d'Herelle) is that the reaction is accidental and one that does not play a part in the activities of normal cultures. When it occurs the bacteria react only passively. Against the nefarious outside influence they attempt to defend themselves; and, if only slightly incapacitated by the "disease," they may eventually recover and, perhaps, even become "immune" to the invading virus. In the alleged mutations that occur, the organisms are merely reacting to a disturbing factor in their environment, and the transformations are never spontaneous.

A second view (that of Bordet) is that, although the reaction is one that controls the trend of normal culture development and tends to keep the culture "true to type," the process is a self-regulating one, controlled by certain products, of the nature of lysins, arising in the bacterial cells. Like other mechanisms of physiologic control, however, it is subject to incapacitation by the force of external environment; and then it can become a mechanism of cell destruction. For Bordet, the reaction in the normal culture is active, while under artificial conditions it may be passive.

In the third place, we can conceive of the bacteriophagic reaction as representing one of the mechanisms for maintaining the wide range of cell and culture changes which are presumably adaptive, which are a constituent part of the specific cyclogeny and which are, therefore, inherent in the biology of the species. Under the influence of unfavorable environment or of artificial treatment this process also may be developed to pathologic proportions. As a result, old culture forms are made to vanish, but the stream of specific germ plasm can never be destroyed. The most fundamental aspect of the bacteriophagic reaction is not lysis and destruction but cell transformation.

13. GENERAL CONCLUSIONS

Having now surveyed the wide field of recent experiment and observation dealing with the Twort-d'Herelle phenomenon, we may bring together in a few words what I regard as the present status of the problem, particularly with reference to the chief theories that have been proposed.

The early theory of Kabeshima has not been supported by any recent evidence. It is clear that the lytic agent does not lie in a source outside the bacteria themselves; moreover, there is not any evidence of its pro-enzyme nature. In all probability, the pro-enzyme of Kabeshima finds its explanation in the now recognized "inciting agents" which may instigate the elaboration of the lytic principle by the bacteria themselves. We must clearly differentiate between the "incitant" and the actually transmissible principle.

While the theory of Bordet and Ciuca has had the greatest number of adherents, neither this theory nor any other based on the lytic action of lysins generated by the cells can be accepted in the light of present knowledge. Recent evidence establishes the probability that the principle is a living rather than a nonliving factor; that it possesses a corpuscular form rather than existing as a substance in solution; that its most significant action is to transform cells, not to destroy them.

Regarding d'Herelle's virus hypothesis, which has experienced the most elaborate development of any of the chief theories, it is clear that the number of adherents is rapidly diminishing. Against it may be brought the following criticisms:

It does not satisfactorily explain the fact that lysis does not occur in physiologic salt solution, or that regeneration of the agent is dependent on a certain degree of multiplication of the bacteria of the substratum.

It does not explain why the cyclostage of the cells from which the bacteriophage seems to be liberated is different from the cyclostage of the cells in which the "invasion" occurs.

It cannot explain the circumstance that the course of cell transformations occurring under the influence of the lytic agent is paralleled (though more slowly) by the transformations occurring in normal enforced, or in spontaneous, dissociation.

It cannot explain the fully established, spontaneous generation of the principle in "normal" sensitive cultures of many bacterial species; nor the rapid generation of the principle from cultures to which have been added only a sterile aqueous solution of pancreatin, trypsin or peptone.

It cannot explain the fact that the bacteriophage can be generated merely by aging sensitive broth cultures for fourteen days in the cold room at a temperature of from 8 to 10 C.

It cannot explain the resistance of the secondary cultures as a result of the acquisition of "immunity" to the principle, for similar forms are developed in all dissociating cultures.

It cannot explain the dual nature of the lytic principle, and the existence of the alpha and beta units which differ markedly, not only in their degree of heat resistance but also in the type of lytic area that each one produces; nor can it explain the specific relation of these fractions of the complete principle to the S and R type cultures; nor the serological specificity of the antilytic serums obtained by the use of these two fractions of the complete bacteriophage.

These points summarize the chief failures of d'Herelle's virus hypothesis; some represent obstacles in other hypotheses also, including my own. But there is still the task of explaining many points which none the less substantiate his view that the agent is endowed with at least some of the attributes of living stuff. If the principle is not a living filtrable virus in the d'Herelle sense, and if it is not the minute "spore form" of a myxomycetean parasite as believed by Kuhn and by Koch and Ziegenspeck, what, then, might it be? Fortunately, the door is not closed to another biologic interpretation.

The working hypothesis—for at present I do not regard it more than that—which I have accepted, is that the bacteriophage is either a definite stage in the cyclogeny of the bacterial species, or a functionally active particle accessory to one of these stages; and by the term "accessory" I mean possessing complementary or reciprocal biologic significance, such, for example, as the relation of sperm cell to ovum. With such a conception there is not any priority of significance in the relation between bacteriophagic corpuscle and the cell that generates it; or, indeed, between the bacteriophagic corpuscle and the cell which it "attacks." Both elements are necessary components of a definite reproductive mechanism possessed by many, if not by all, bacteria This constitutes the nucleus of what I have termed my "homogamic theory" of bacteriophage action.

Of the actual existence of such a mechanism in the reproductive processes of bacteria, unfortunately there is, as yet, no valid proof except in the still somewhat obscure observations presented by Enderlein, dealing with assumed sexual reproduction in the cholera vibrio. Regarding this point, however, one must bear in mind that the essential cytologic results of Enderlein have been confirmed in a large measure by the observations of Schumacher [368]; and, it should be added, without previous knowledge of Enderlein's earlier work. Moreover, for one who has followed the recent trend of studies on microbic dissociation, the observations, if not the final conclusions, of P. Kuhn cannot fail to possess the greatest significance. The things he has observed can safely be regarded as inherent in the dissociative reaction and, though still lacking in the intimate details, tend to afford valuable support to the homogamic theory, which, I believe, offers the most logical working basis yet provided for further study of the fascinating, but highly complex, problem.

In closing this critical, but also, I hope, somewhat constructive, review, I merely wish to add that no one appreciates more than I do the careful and persevering experimentation given to the study of the bacteriophage phenomenon by d'Herelle. Moreover, it cannot for a moment be doubted that the majority of his observations possess a lasting value; nor that the failure of workers to recognize the bacteriophage as a parasitizing virus of bacteria will detract from the achievements in medicine which his work has already made possible. Taken as a whole, the grounds for present controversy lie not so much in the recognition of established facts as in their interpretation. And modes of interpretation in matters of this sort have, unfortunately, been largely determined in advance by one's biologic perspective on what I regard at present, as the most fundamental of all bacteriologic problems— the question of monomorphism and the reproductive behavior of bacteria. The downfall of monomorphism carries with it, not only the virus theory, but also all other current theories of the lytic phenomenon; and it may be confidently hoped that the elaboration and support of such theories represent the last of the biologic errors that can be laid at the

door of this false conception of the nature of bacteria and their reproductive mechanism.   The future views, not of transmissible autolysis
alone, but of several other highly important aspects of bacterial behavior.
must, it seems to me, be rebuilt on a new and firmer foundation—on
the clear recognition of the far-reaching significance of bacterial
cyclogeny and the often curious, dissociative reactions which are its
natural and inevitable consequents.

REFERENCES

[1] Allesandrini, A.: Sulla natura del batteriofargo, Ann. d'hyg., 1925, p. 649; referred to
in Centralbl. f. Bakteriol., I, Ref., 1926, 83, p. 75.
[2] Almquist, E.: Investigations on bacterial hybrids, J. Infect. Dis., 1924, 35, p. 75.
[3] Almquist, E.: Biologische Forschungen über die Bakterien, O. Weigel, 1925, Leipzig.
[4] Alphonsi, M.: Guérison rapide de deux cas de pyelonéphrite gravidique traités par le
[5] Von Angerer: Beitrage zum Bakteriophagen-Problem, Arch. f. Hyg., 1924, 92, p. 312.
bactériophage de d'Hérelle, Bull. Soc. d'bst. et de gynéc. de Paris, 1924, 13, p. 341.
[6] Applemans, R.: Quelques applications de la méthode de dosage du bactériophage, Compt.
rend. Soc. de biol., 1922, 86, p. 508.
[7] Applemans, R.: Le dosage de bactériophage, Compt. rend. Soc. de biol., 1921, 85, p. 1098.
[8] Arkwright, J. A.: Variations in bacteria in relation to agglutination both by salts and by
specific serum, J. Path. & Bact., 1921, 24, p. 36.
[9] Arkwright, J. A.: The source and characteristics of certain cultures sensitive to the
bacteriophage, Brit. J. Exper. Path., 1924, 5, p. 23.
[10] Arloing, F., and Langeron, L.: Action de l'eau de Javel sur diversés éspecès de bactériophage, Compt. rend. Soc. de biol., 1927, 96, p. 454.
[11] Arloing, F.; Langeron, L., and Sempé: Recherches comparatives sur l'action de certains
agents physiques sur le bactériophage, un diastase et le complément, Compt. rend. Soc. de
biol., 1925, 92, p. 260.
[12] Arloing, F., and Sempé: Pouvoir antimicrobienne lytique d'eaux fluviales ou marines,
français et étrangères; rôle possible du bactériophage, Compt. rend. Soc. de biol., 1926, 94, p. 191.
[13] Arloing, F.; Sempé and Chavanne: Propriétés anti-microbienne de diversés eaux fluviales
ou marines; pouvoir bactériophagique, Bull. Acad. de méd., Paris, 1925, 93, p. 184.
[13a] Arloing, F., and Chavanne: De l'influence sur le bactériophage des électrolytes et de la
concentration en ions H du milieu, Compt. rend. Soc. de biol., 1925, 93, p. 531.
[14] Arnold, L.: Bacteriophage phenomena, J. Lab. & Clin. Med., 1923, 8, p. 720.
[14a] Arnold, L.: The significance of the bacteriophage in surface water, J. Am. Pub. Health
A., 1925, 15, p. 950.
    → Arnold, L., and Weiss, E.: A study of bacteriophage with antibacteriophage serum,
J. Infect. Dis., 1924, 35, p. 505.
[15] Arnold, L., and Weiss, E.: Isolation of bacteriophage from bacterial proteins, J. Infect.
Dis., 1925, 37, p. 411.
[16] Arnold, L., and Weiss, E.: Bacterial protein-free bacteriophage prepared by Tryptic
digestion, J. Immunol., 1926, 12, p. 393.
[16a] Arnold, L., and Weiss, E.: Prophylactic and therapeutic possibilities of the Twort-
d'Herelle's bacteriophage, J. Lab. & Clin. Med., 1926, 12, p. 20.
[17] Asheshov, I.: Experimental studies on the bacteriophage, J. Infect. Dis., 1924, 34, p. 536.
[17a] Asheshov, I.: Quelques récherches sur la nature des plaques de bactériophage, Compt.
rend. Soc. de biol., 1923, 89, p. 120.
[17b] Asheshov, I.: Le pouvoir antigène des lysats ultrasteriles, Compt. rend. Soc. de biol.,
1925, 93, pp. 643, 644.
[18] Asheshov, I.: Action de citrate de soude sur le bactériophage, Compt. rend. Soc. de biol.,
1926, 94, p. 687.
[19] Asheshova, Inna: Bactériophage du bacille pyocyanique, Compt. rend Soc. de biol.,
1926, 95, p. 1029.
[20] Bablet, J.: Sur le principe bactériophage de d'Hérelle, Compt. rend. Soc. de biol., 1920,
83, p. 1322.
[21] Bachmann, A., and Aquino, L.-I.: Sur le bactériophage, Compt. rend Soc. de biol.,
1922, 86, p. 1108.
[22] Bachmann, A., and de la Barrera: Quelques variations sérologique du bacille paratyphique,
Compt. rend Soc. de biol., 1923, 89, p. 56.
[23] Bagger, S. V.: The enterococcus, J. Path. & Bact., 1926, 29, p. 225.
[24] Bail, O.: Theorie der Bakteriophagenwirkung, Bull. tech. de Sc. méd. Genève, 1925,
1, p. 23. Cited after d'Herelle.[103]
[25] Bail, O.: Der Stand und die Ergebnisse der Bakteriophagenforschung, Deutsche med.
Wchnschr., 1925, 51, p. 13; referred to in Centralbl. f. Bakteriol., 1, ref., 1925, 78, p. 513.

[26] Bail, O.: Elementarbakteriophagen des Shigabazillus, Wien. klin. Wchnschr., 1922, 35, pp. 722, 743 and 765.

[26a] Bail, O.: Bakteriophagen Wirkung gegen Flexner- und Coli-Bakterien, Wien. klin. Wchnschr., 1921, 34, p. 448.

[27] Bail, O., and Watanabe: Versuche ueber specifische Bakteriophagenwirkung, Wien. klin. Wchnschr., 1922, 35, pp. 8 and 362.

[28] Beckerish, A., and Hauduroy, P.: Le bactériophage dans le traitement de la fièvre typhöid, Compt. rend. Soc. de biol., 1922, 86, p. 168.

[28a] Beckerish, A., and Hauduroy, P.: Le traitement des infections urinaires à colibacilles par le bactériophage de d'Hérelle, Bull. méd., 1923, 37, p. 273.

[29] Beckerish, A., and Hauduroy, P.: Le bactériophage de d'Hérelle: Ses applications thérapeutiques, J. Bact., 1923, 8, p. 163.

[30] Beckerish, A., and Hauduroy, P.: Au sujet de l'obtention de bactériophage par antagonisme microbienne, Compt. rend. Soc. de biol., 1922, 86, p. 881.

[31] Béguet, M.: Hypothèses sur le rôle de la pression osmotique dans les phénomenès microbiens: I. phénomenè de d'Hérelle; II. variations et mutations des types microbiennes, Arch. d' inst. Pasteur de l'Afrique du Nord, 1927, 5, p. 25.

[32] Bergstrand, H.: Sur la lyse microbienne transmissible, Compt. rend. Soc. de biol., 1922, 86, p. 489.

[33] Biemond, A. G.: Einige Bakteriophagen Untersuchungen, Ztschr. f. Hyg., 1924, 103, p. 681.

[34] Borchardt, W.: Weitere biologische Beiträge zum d'Herelleschen Phänomen. Klin. Wchnschr., 1923, 2, pp. 295 and 791.

[35] Borchardt, W.: Biologische Beiträge zum d'Herelleschen Phänomen, Ztschr. f. Immunitätsforsch. u. exper. Therap., 1923, 37, p. 1.

[36] Bordet, J.: Obtention de principes de faible puissance dans l'autolyse microbienne transmissible, Compt. rend. Soc. de biol., 1922, 87, p. 987.

[37] Bordet, J.: Le rôle des sels de calcium dans l'autolysis microbienne transmissible, Compt. rend. Soc. de biol., 1926, 94, p. 403.

[38] Bordet, J.: La théorie de l'autolysis transmissible et les objections de d'Hérelle, Compt. rend. Soc. de biol., 1925, 93, p. 1432.

[39] Bordet, J.: Pouvoir lysogène actif ou spontané et pouvoir lysogène passif ou provoqué, Compt. rend. Soc. de biol., 1925, 93, p. 1054.

[40] Bordet, J.: De l'autolyse microbienne transmissible ou du bactériophage, Ann. de l'Inst. Pasteur. 1925, 39, p. 717.

[41] Bordet, J., and Ciuca, M.: Exudates leucocytaire et autolysis microbienne transmissible, Compt. rend. Soc. de biol., 1920, 83, p. 1293.

[42] Bordet, J., and Ciuca, M.: La bactériophage de d'Hérelle; sa production et son interprétation, Compt. rend. Soc. de biol., 1920, 83, p. 1296.

[43] Bordet, J., and Ciuca, M.: Autolysis microbienne et sérum antilytique, Compt. rend. Soc. de biol., 1921, 84, p. 280.

[44] Bordet, J., and Ciuca, M.: Spécificité de l'autolyse microbienne transmissible, Compt. rend. Soc. de biol., 1921, 84, p. 278.

[45] Bordet, J., and Ciuca, M.: Sur la régénération du principle actif dans l'autolyse microbienne, Compt. rend. Soc. de biol., 1921, 85, p. 1095.

[46] Bordet, J., and Ciuca, M.: Variations d'energie du principe actif dans l'autolyse microbienne transmissible, Compt. rend. Soc. de biol., 1922, 87, p. 366.

[47] Bordet, J., and Ciuca, M.: Autolyse microbienne et sérum antilytique, Compt. rend. Soc. de biol., 1921, 84, p. 280.

[48] Botez, A.: La bactériolyse en série par le violet de méthyle, Compt. rend. Soc. de biol., 1921, 85, p. 585.

[49] Boulet, P.: Le bactériophage de d'Hérelle existe-il? Thèse de Montpéllier, L'Abeille, 1924.

[50] Breinl, F., and Fischer, M.: Variationserscheinungen in der Paratyphusgruppe, Ztschr. f. Immunitätsforsch. u. exper. Therap., 1923, 35, p. 205.

[51] Breinl, F., and Hoder, F.: Bakteriophagenwirkung in Paratyphusgruppe, Centralbl. f. Bakteriol., 1, O., 1925, 96, p. 1.

[52] Bronfenbrenner, J.: Further evidence of the resistance of the bacteriophage to Alcohol, Proc. Soc. Exper. Biol. & Med., 1927, 24, p. 372.

[53] Bronfenbrenner, J., and Korb, C.: Studies on the bacteriophage of d'Herelle: II. Effects of Alcohol on Bacteriophage, J. Exper. Med., 1925, 42, p. 419.

[54] Bronfenbrenner, J., and Korb, C.: Studies on bacteriophage: III. Some of the factors determining the size of plaques of bacterial lysis on agar, J. Exper. Med., 1925, 42, p. 483.

[55] Bronfenbrenner, J., and Korb, C.: Studies on the bacteriophage: IV. Concerning the oneness of the bacteriophage, J. Exper. Med., 1925, 42, p. 821.

[56] Bronfenbrenner, J., and Korb, C.: Studies on the bacteriophage: V. Effect of Electrolytes on the rate of inactivation by alcohol, J. Exper. Med., 1926, 43, p. 71.

[57] Bronfenbrenner, J., and Korb, C.: On variants of B. pestis caviae resistant to lysis by the bacteriophage, Proc. Soc. Exper. Biol. & Med., 1925, 23, p. 3.

[58] Bronfenbrenner, J., and Muckenfuss, R.: The lysis of dead bacteria by bacteriophage, Proc. Soc. Exper. Biol. & Med., 1926, 23, p. 633.

[59] Bronfenbrenner, J.; Muckenfuss, R. and Korb, R.: Studies on bacteriophage: VI. On the Virulence of the overgrowth in the lysed cultures of B. pestis caviae (M. T. II), J. Exper. Med., 1926, 44, p. 605.

[60] Brown, J. H., and Basaca, M.: Pseudobacteriophage of B. anthracis, Proc. Soc. Exper. Biol. & Med., 1926, 23, p. 625.

[61] Bruce-White, P. B.: Serological studies with regard to the classification of the Salmonella group, Special Report no. 91, M. Research Council, 1925.

[62] Bruce-White, P. B.: Special Report no. 103, M. Research Council, 1926.

[63] Brutsaert, P.: Le bactériophage dans les milieux gélatinés, Compt. rend. Soc. de biol., 1924, 90, p. 1292.

[64] Brutsaert, P.: Les bactériophages et les microbes dans le boullion hypersale, Compt. rend. Soc. de biol., 1924, 90, p. 646.

[65] Brutsaert, P.: L'agglutination des microbes résistants, Compt. rend. Soc. de biol., 1924, 90, p. 645.

[66] Bruynoghe, R.: Au sujet de la guérison des germes devenus résistants au principe bactériophage, Compt. rend. Soc. de biol., 1921, 85, p. 20.

[67] Bruynoghe, R.: De la nature des principe bactériophage, Compt. rend. Soc. de biol., 1921, 85, p. 258.

[68] Bruynoghe, R., and Dubois, A.: La précipitation spécifique des bactériophages, Compt. rend. Soc. de biol., 1927, 96, p. 211.

[69] Bruynoghe, R., and Dubois, A.: La parenté des microbes devenus résistants au bactériophage, Compt. rend. Soc. de biol., 1927, 96, p. 209.

[70] Bruynoghe, R., and Dubois: Sur une substance inhibitive produite par un souche de Bacille typhosus, Compt. rend. Soc. de biol., 1927, 96. p. 429.

[71] Bruynoghe, R., and Maisin, J.: Au sujet de l'unite du principe bactériophage, Compt. rend. Soc. de biol., 1921, 85, p. 1122.

[72] Bruynoghe, R., and Maisin, J.: Essais de thérapeutique au moyen du bactériophage staphylocoque, Compt. rend. Soc. de biol., 1921, 85, p. 1120.

[73] Bruynoghe, R., and Maisin, J.: Le principe bactériophage du staphylocoque, Compt. rend. Soc. de biol., 1921, 85, p. 1118.

[74] Bruynoghe, R., and Mund, W.: Les microbes irradiés et le bactériophage, Compt. rend. Soc. de biol., 1925, 92, p. 464.

[75] Bruynoghe, R., and Wagemans: La résistance des bactériophages au sérum neutralissant, Compt. rend. Soc. de biol., 1923, 88, p. 968.

[76] Burnet, E.: Actions d'entrainement entre races et éspéces microbienne, Arch. de l'Inst. Pasteur (Tunis), 1925, 14, p. 34.

[77] Burnett, F. M.: The nature of acquired resistance to bacteriophage, J. Path. & Bact., 1925, 28, p. 407.

[78] Burnet, F. M.: The relationships between heat-stable agglutinogens and sensitivity to bacteriophage in the Salmonella group, Brit. J. Exper. Path., 1927, 8, p. 121.

[79] Busson, B., and Ogata, N.: Untersuchungen über sekundare und bakteriophagenresistente Dysenterie-stämme und ihre Beziehung zu dem sog; Schmitzstämmen, Wien. klin. Wchnschr., 1924, 37, p. 665; referred to in Centralbl. f. Bakteriol., 1925, 78, p. 40.

[80] Caldwell, J.: The activity of an anticolon bacteriophage in synthetic medium, J. Infect. Dis., 1926, 39, p. 122.

[81] Caldwell, J.: Sewage filtrate as a source of bacteriophage, J. Infect. Dis., 1927, 40, p. 575.

[82] Callow, B. R.: Bacteriophage phenomena with Staphylococcus aureus, J. Infect. Dis., 1922, 30, p. 643.

[83] Callow, B. R.: Further studies on staphylococcus bacteriophage, J. Infect. Dis., 1927, 41, p. 124.

[84] Čanzik, J.: Bacteriophagy in Pyocyaneus Cultures, Čas. lék. Česk., 1923, 62, p. 25.

[85] Carrel, A.: La genèse des sarcomes, Compt. rend. Soc. de biol., 1925, 93, p. 1491.

[86] Caublot, P.: Le bactériophage du pneumobacille de Friedländer, Compt. rend. Soc. de biol., 1924, 90, p. 622.

[87] Ciuca, M.: Lyse transmissible en absence d'électrolytes libres, Compt. rend. Soc. de biol., 1924, 90, p. 521.

[88] Ciuca, M., and Manoliu, E.: Action inhibitrice du filtrat de cultures et lyse transmissible au cours de la fiévre typhoide, Compt. rend. Soc. de biol., 1924, 91, p. 1225.

[89] Clark, P. F., and Clark, A. S.: A bacteriophage agent against a virulent, hemolytic streptococcus, Proc. Soc. Exper. Biol. & Med., 1927, 24, p. 635.

[90] Collins, George: Studies on the source of the bacteriophage and on the origin of transmissible bacterial autolysis, 1924.

[91] Combiesco, D.: Sur le phénomène de d'Hérelle, Compt. rend. Soc. de biol., 1922, 87, 17.

[92] Coons, G. H., and Kotilia, J. E.: The transmissible lytic principle (bacteriophage) in relation to plant pathogens, Phytopathology, 1925, 15, p. 357.

[93] Da Costa Cruz, J.: Sur la nature du bactériophage, Compt. rend. Soc. de biol., 1923, 89, p. 959.

[94] Da Costa Cruz, J.: Sur la nature du bactériophage. À propos d'une note de d'Hérelle, Compt. rend Soc. de biol., 1924, 90, p. 694.

[95] Da Costa Cruz, J.: Sur le méchanism de l'action anti-lytique du sérum antibactérien dans la lyse par le bactériophage, Compt. rend. Soc. de biol., 1924, 91, p. 840.

[96] Da Costa Cruz, J.: Le traitement des dysenteries bacillaires par le bactériophage, Compt. rend Soc. de biol., 1924, 91, p. 845.

[97] Da Costa Cruz, J.: Sur l'influence des électrolytes dans la lyse par le bactériophage, Compt. rend. Soc. de biol., 1924, 90, pp. 236-694.

[98] Da Costa Cruz, J.: Action du sérum anti-bacterien dans la lyse par le bactériophage, Compt. rend. Soc. de biol., 1926, 95, p. 1437.

[99] Da Costa Cruz, J.: La lyse par le bactériophage observée au microscope, Compt. rend. Soc. de biol., 1926, 95, p. 1501.

[100] Da Costa Cruz, J.: Action anti-lytique du sérum anti-bacterien dans la lyse par le bactériophage, Compt. rend. Soc. de biol., 1926, 95, p. 1006.

[101] Da Costa Cruz, J.: Pouvoir lysogène spontané du Bacillus coli de Lisbonne et Carrère, Compt. rend. Soc. de biol., 1927, 97, p. 837.

[102] Courcoux, A.: Pregnancy pyelonéphrite, Bull. et mém. Soc. méd. d. hóp de Paris, 1922, 46, p. 1151; cited after Larkum (footnote 227).

[103] Cowie, D. M.: Observations on the bacteriophage, Ann. Clin. Med., 1926, 5, p. 57.

[104] Davison, W. C.: Observations on the nature of the bacteriophage, J. Bact., 1922, 7, p. 491.

[105] Delsase, R.: Le bactériophage de d'Hérelle; ses applications en thérapeutique urinaire, Presse méd., 1926, 83, p. 458.

[106] Dewey, E., and Green, R.: Bacteriophages in bacterial cultures, Proc. Soc. Exper. Biol. & Med., 1927, 24, p. 911.

[107] Dimtza, A.: Ueber Veränderungen von Colistämmen durch Bakteriophagenwirkung "in vivo et in vitro," Centralbl. f. Bakteriol., 1, O., 1926, 101, p. 171.

[108] Doerr, R.: Die Bakteriophagen (Phänomen von Twort und d'Herelle, Klin. Wchnschr., 1922, 1, pp. 1489 and 1537; cited after d'Herelle.

[108a] Doerr, R.: Die invisiblen Ansteckungsstoffe und ihre Beziehungen zu Problemen der allgemeinen Biologie, Klin. Wchnschr., 1923, 2, p. 209.

[109] Dumas, J.: Sur la présence du bactériophage dans l'intestin sain, dans la terre et dans l'eau, Compt. rend. Soc. de biol., 1920, 83, p. 1314.

[110] Dutton, L. O.: Rôle of the bacteriophage in streptococcus infections: an interpretation of certain cultural characteristics, J. Infect. Dis., 1926, 39, p. 48.

[111] Dutton, L. O.: The probable rôle of the bacteriophage in streptococcus infections, J. Lab. & Clin. Med., 1926, 11, p. 763.

[112] Eastwood, A.: Bacterial variation and transmissible autolysis, Report no. 18 on Public Health and Medical Subjects, Ministry of Health, 1923.

[113] Eisenberg, P.: Ueber die Anpassung der Bakteria an die Abwehrkrafte des infizierten Organismus, Centralbl. f. Bakteriol., 1, O., 1903, 34, p. 739.

[114] Elder, A. L., and Tanner, F. W.: Action of the Bacteriophage on a Low Temperature Organism, Proc. Soc. Exper. Biol. & Med., 1927, 24, p. 645.

[115] Eliava, G., and Suarez, E.: Au sujet de l'ultrafiltration du corpuscle bactériophage, Compt. rend. Soc. de biol., 1927, 96, p. 460.

[116] Enderlein, G.: Grundelementeder vergleichenden Morphologie und Biologie der Bakterien (Bakteriologische Studien III), 1916, p. 403.

[117] Enderlein, G.: Bakterien-Cyclogenie, Prologomena zu Untersuchungen über Bau, geschlechtliche und ungesschlechtliche Fortpflanzung und Entwicklung der Bakterien, 1925, p. 1.

[118] Epstein, T., and Fejgin, B.: Su la résistance des staphylococques hémolysants vis-a-vis du bactériophage spécifique, Compt. rend. Sos. de biol., 1926, 95, p. 908.

[119] Euguchi, C.: Studien über das d'Herellesche Phänomen, über Dysenterie-bakteriophagen, Sarkingaku Zashi, 1923, no. 332; referred to in Centralbl. f. Bakteriol., 1, ref., 1925, 78, p. 40.

[120] Farby, P.: Autolysemicrobienne transmissible obtenué par antagonisme microbien, Compt. rend. Soc. de biol., 1922, 87, p. 369.

[121] Farby, P., and van Beneden, J.: Sérum antilytique at sérum antilytique, Compt. rend. Soc. de biol., 1924, 90, p. 111.

[122] Farby, P., and van Beneden, J.: À propos de obtention de l'autolysis transmissible par antagonisme, Compt. rend. Soc. de biol., 1924, 90, p. 109.

[123] Fejgin, B.: Sur la form filtrant de Bacille d'Éberth, Compt. rend. Soc. de biol., 1925, 92, p. 1528.

[124] Fejgin, B.: Contribution a l'étude des races résistantes du bacille de Shiga-Kruse, Compt. rend. Soc. de biol., 1923, 90, p. 1381.

[125] Fejgin, B.: Sur les cultures secondaires du bacille typhique isolés des cobayes infectés avec le virus de la fièvre typhoïde, Compt. rend. Soc. de biol., 1925, 93, p. 1530.

[126] Fejgin, B.: Sur les variations brusques du Proteus HX19 survenues sous l'influence de l'agent lytique anti HX19, et leur rapport avec les souches isolés des cobayes infectés avec le virus de passage du typhus exanthématique, Compt. rend. Soc. de biol., 1924, 90, p. 1106.

[127] Fejgin, B.: Sur le principe lytique diphthérétique, Compt. rend. Soc. de biol., 1925, 93, p. 365.

[128] Fejgin, B., and Lararevitz, M.: A propos des bactéries a partier du Proteus X19 lyse, Compt. rend. Soc. de biol., 1927, 96, p. 342.

[129] Fejgin, B., and Supeniewski: Sur la nature du phénomène de d'Hérelle, Compt. rend. Soc. de biol., 1923, 89, p. 1385.

[130] Fleming, A.: On a remarkable bacteriolytic element found in tissues and secretions, Proc. Royal Soc., London, series B, 1922-1923, 94, p. 306; Fleming and Allison: Brit. J. Exper. Path., 1922, 3, p. 252.

[131] Flu, P. C.: Das Verhalten eines inagglutinablen Flexnerbakterien gegenüber Anti-flexnerbakteriophagen, Centralbl. f. Bakteriol., 1, O., 1923, 90, p. 374.

[132] Flu, P. C.: Ueber Cholerabakteriophagen, Tijdschr. v. vergelijk. Geneesk., 1924, 10, p. 196.

[133] Flu, P. C.: Komplementbindungsversuche mit Kannichenserum gegenüber Bakteriophagen un Bakterienextrakten, Centralbl. f. Bakteriol., 1, O., 1926, 83, p. 76.

[134] Flu, P. C.: Sur la nature du bactériophage, Compt. rend. Soc. de biol., 1927, 96, p. 1148.

[135] Gaté, J., and Gardère, H.: Apparition tardive d'un principe bactériophage de deux souches de bacilles pyocyanique au cours de leur vie cultivable, Compt. rend. Soc. de biol., 1927, 96, p. 545.

[136] Gerretsen, F.; Gryns, A.; Sack, J., and Söhnigen, N.: Das Vorkommen eines Bakteriophagen in dem Wurzelknollchen der Leguminosen, Centralbl. f. Bakteriol., 2, O., 60, p. 311.

[137] Gildemeister, E.: Ueber das d'Herelleschen Phänomen, Berl. klin. Wchnschr., 1921, 58, p. 1355.

[137a] Gildemeister, E.: Weitere Untersuchungen über das d'Herelleschen Phänomen, Centralbl. f. Bakteriol., 1, O., 1922, 89, p. 181.

[138] Gildemeister, E., and Herzberg, K.: Zur Theorie der Bakteriophagen (d'Herelle' Lysine), Centralbl. f. Bakteriol., 1924, 93, p. 402; ibid., 1923, 91, p. 402.

[139] Gjorup, E.: Investigations on d'Herelle's phenomenon, Kopenhagen, 1925; referred to in Centralbl. f. Bakteriol., 1926, 82, p. 280.

[140] Gohs, W.: Eine Theorie der Bakteriophagenwirkung und ihre Beziehung zu Immunität, Anaphylaxie und Verdauung, Ztschr. f. Immuntätsforsch. u. exper. Therap., 1925, 45, pp. 141, 269, 413; referred to in Centralbl. f. Bakteriol., 1926, 82, p. 279.

[141] Gohs, W.: Eine neue Methode des Nachweiss des bakteriophagen Lysins, Ztschr. f. Immunitätsforsch. u. exper. Therap., 1926, 49, p. 139.

[142] Gohs, W.: Eine neue Theorie der Bakteriophagenwirkung und ihre Beziehung zu Immunität Anaphylaxie und Verdauung, Ztschr. f. Immunitätsforsch. u. exper. Therap., 1927, 49, p. 532.

[143] Gohs, W., and Jacobsohn, Irene: Ueber die Lysoresistenz und Lysofähigkeit der sekundären Kulturen beim d'Herelleschen Phänomen, Ztschr. f. Immunitätsforsch. u. exper. Therap., 1926, 49, p. 17.

[144] Gohs, W., and Jacobsohn, Irene: Ueber die Bindung des Bakteriophagen Lysins durch die Bakterien, Ztschr. f. Immuntätsforsch. u. exper. Therap., 1926, 49, p. 412.

[145] Goyle, A. N.: On bacterial variation with special reference to the alleged convergence phenomenon exhibited by certain distinct pathogenic species, J. Path. & Bact., 1926, 29, p. 149.

[146] Gougerot and Peyre, E.: Le bactériophage dans le traitement des affections cutaneés, Compt. rend. Soc. de biol., 1924, 91, p. 452.

[147] Grasset, E.: Recherches sur le passage du bactériophage à travers le placenta, Compt. rend Soc. de biol., 1927, 96, p. 839.

[148] Gratia, A.: The Twort-d'Herelle Phenomenon, J. Exper. Med., 1921, 35, pp. 115, 287.

[149] Gratia, A.: de l'adaptation héréditaire du coli bacille à l'autolyse microbienne transmissible, Compt. rend. Soc. de biol., 1921, 84, p. 750.

[150] Gratia, A.: Autolysis transmissible et variations microbiennes, Compt. rend. Soc. de biol., 1921, 85, p. 251.

[151] Gratia, A.: La lyse transmissible du staphylocoque et ses applications thérapeutique, Bull. de l'Acad. roy de méd. de Belgique, 1922, 2, p. 72.

[152] Gratia, A.: La lyse transmissible du staphylocoque; sa production ses applications thérapeutiques, Compt. rend. Soc. de biol., 1922, 86, p. 276.

[153] Gratia, A.: Preliminary report on a staphylococcus bacteriophage, Proc. Soc. Exper. Biol. & Med., 1921, 18, p. 217.

[154] Gratia, A.: L'autolyse transmissible du staphylocoque et l'action coagulante des cultures lyseés, Compt. rend. Soc. de biol., 1921, 85, p. 25.

[155] Gratia, A.: Hétérogéneité du principe lytique du colibacille, Compt. rend. Soc. de biol., 1923, 89, p. 821.

[156] Gratia, A.: Rélation entre la variabilité du coli bacille et l'hétérogéneité du principe lytique correspondant, Compt. rend. Soc. de biol., 1923, 89, p. 824.

[157] Gratia, A.: Sur un remarkable example d'antagonisme entre deux souches de coli bacille, Compt. rend. Soc. de biol., 1925, 93, p. 1040.

[158] Gratia, A., and Dath, Sarah: Propriétés bactériolytique de certaines moisissures, Compt. rend. Soc. de biol., 1924, 91, p. 1442.

[159] Gratia, A., and Dath, Sarah: Moisissures et microbes bactériophages, Compt. rend. Soc. de biol., 1925, 92, p. 461.

[160] Gratia, A., and Dath, Sarah: Propriétés bactériolytiques des streptothrix, Compt. rend. Soc. de biol., 1926, 94, p. 1267.

[161] Gratia, A., and de Kruif, Lois: Tentative d'isolement de bactériophages d'inegale activité à partir d'une principe lytique coli manifestant des variations d'energie, Compt. rend. Soc. de biol., 1923, 88, p. 629.

[161a] Gratia, A., and Rhodes, B.: Action du principe lytique sur les émulsions de staphylocoques vivant et de staphylocoques tués, Compt. rend. Soc. de biol., 1923, 89, p. 1171; 1923, 90, p. 640.

[162] Gregorieff, A.: Sur le mode d'inactivation des bactériophages, Compt. rend. Soc. de biol., 1927, 96, p. 1141.

[163] Hadley, P.: Transmissible lysis in Bacillus pyocyaneus, J. Infect. Dis., 1924, 84, p. 260.

[164] Hadley, P.: The action of the lytic principle on capsulated bacteria, Proc. Soc. Exper. Biol. & Med., 1925, 23, p. 109.

[165] Hadley, P.: The variation in size of lytic areas and its significance, J. Bact., 1924, 9, p. 397.

[166] Hadley, P.: Proliferative reaction to stimuli by the lytic principle (bacteriophage) and its significance, J. Infect. Dis., 1925, 37, p. 35.

[167] Hadley, P.: Parallelism between serologic and bacteriophagic response in B. typhosus and certain avian paratyphoids, Proc. Soc. Exper. Biol. & Med., 1926, 23, p. 443.

→ Hadley, P.: Microbic dissociation: the instability of bacterial species with special reference to active dissociation and transmissible autolysis, J. Infect. Dis., 1927, 40, p. 1.

[168a] Hadley:,[168] p. 40.

[168b] Hadley,[168] p. 248.

[168c] Hadley,[168] p. 255.

[168d] Hadley,[168] p. 256.

[168e] Hadley,[168] p. 244.

[168f] Hadley,[168] p. 272.

[169] Hadley, P., and Dabney, Eugenia: The bacteriophagic relationships between B. coli, S. fecalis and S. lacticus, Proc. Soc. Exper. Biol. & Med., 1926, 24, p. 13.

[169a] Hadley, Philip; and Dabney, Eugenia: The dual nature of the lytic principle: A study of the alpha and beta units of an anti-Paratyphoid bacteriophage, Proc. Soc. Exp. Biol. and Med., 1928, 25, p. 355.

[170] Hadley, P., and Klimek, John: The so-called "origin" or "source" of the bacteriophage, Proc. Soc. Exper. Biol. & Med., 1927, 25, p. 34.

[171] Hauduroy, P.: De l'action du sérum anti-dysentérique sur la lyse du bacille de Shiga par le bactériophage de d'Hérelle, Compt. rend. Soc. de biol., 1922, 87, p. 966.

[172] Hauduroy, P.: Sur les lysines du bactériophage de d'Hérelle, Compt. rend. Soc. de biol., 1922, 87, p. 964.

[173] Hauduroy, P.: Constitution du bactériophage de d'Hérelle, Compt. rend. Soc. de biol., 1923, 88, p. 59.

[174] Hauduroy, P.: Le rôle du bactériophage dans le fièvre typhoïde; sa présence dans le sang, Compt. rend. Soc. de biol., 1923, 89, p. 875.

[175] Hauduroy, P.: Sensibilization d'animaux à certaines infections par une vaccination anti-bactériophage, Compt. rend. Soc. de biol., 1924, 90, p. 290.

[176] Hauduroy, P.: Les cultures secondaires après filtration, dans le phénomène de d'Hérelle, Compt. rend. Soc. de biol., 1924, 91, p. 1209.

[177] Hauduroy, P.: Les cultures secondaires après filtration, dans le phénomène de d'Hérelle: II, Compt. rend. Soc. de biol., 1924, 91, p. 1325.

[178] Hauduroy, P.: Action de gélatine sur le phénomène de d'Hérelle, Compt. rend. Soc. de biol., 1924, 90, p. 1463.

[179] Hauduroy, P.: Le rôle du bactériophage dans la fièvre typhoïde, Compt. rend Soc. de biol., 1925, 93, p. 100.

[180] Hauduroy, P.: Le rôle du bactériophage de d'Hérelle dans la guérison de la fièvre typhoïde, Presse méd., 1925, 32, p. 525.

[181] Hauduroy, P.: Action de la bile sur le bactériophage et importance de cette action, Compte rend. Soc. de biol., 1925, 92, p. 1442.

[182] Hauduroy, P.: Le bactériophage de d'Hérelle, Médecine, 1926, 7, p. 3.

[183] Hauduroy, P.: Traitement des infections à staphylococcus par le bactériophage, Presse méd., 1926, 34, p. 1195.

[184] Hauduroy, P.: Les formes invisible des microbes visible, Rev. de path. comp. et d'hyg. gen., Jan. 20, 1926.

[185] D'Herelle, F.: Sur unmicrobe invisible antagoniste des bacille dysentériques, Compt. rend. Acad. d. sc., 1917, 165, p. 373.

[186] D'Herelle, F.: Technique de la récherche du microbe filtrant bactériophage (Bactériophagum intestinale), Compt. rend. Soc. de biol., 1918, 81, p. 1160.

[187] D'Herelle, F.: L'ultramicrobe bactériophage, Compt. rend. Soc. de biol., 1921, 85, p. 767.

[188] D'Herelle, F.: The bacteriophage: its rôle in immunity, 1922.

[189] D'Herelle, F.: Sur le présence du bactériophage dans les leucocytes, Compt. rend. Soc. de biol., 1922, 86, p. 477.

[190] D'Herelle, F.: Sur un "principe bactériolysant," non bactériophage, existant dans l'intestin des cholériques, Compt. rend. Soc. de biol., 1923, 88, p. 723.

[191] D'Herelle, F.: Sur un "principe bactériolysant" non bactériophage, existant dans l'intestin des cholériques, Compt. rend. Soc. de biol., 1923, 88, p. 72.

[192] D'Herelle, F.: Immunity in the natural infectious diseases, 1924.

[193] D'Herelle, F.: The bacteriophage and its behavior, 1926.

[193a] D'Herelle,[193] p. 71.

[193b] D'Herelle,[193] p. 220.

[193c] D'Herelle,[193] p. 73.

[193d] D'Herelle,[193] p. 326.

[193e] D'Herelle,[193] p. 222.

[194] D'Herelle, F.: La bactériophage et hyperaerobiose, Compt. rend. Soc. de biol., 1927, 96, p. 451.

[195] D'Herelle, F., and Eliava, G.: Sur le sérum anti-bactériophage, Compt. rend. Soc. de biol., 84, p. 719.

[196] D'Herelle, F., and Hauduroy, P.: Sur les charactéres des symbiose "Bactérie-Bactériophage," Compt. rend. Soc. de biol., 1925, 94, p. 1288.

[197] Hoder, F.: Ueber Zusammenhänge zwischen Bakteriophagen und Bakterienmutation, Ztschr. f. Immunitätsforsch. u. exper. Therap., 42, p. 197.

[198] Hoder, F., and Suzuki, K.: Ueber die Gewinnung von Bakteriophagen aus Pancreasextrakten, Centralbl. f. Bakteriol., 1, O., 1926, 98, p. 433.

[199] Ikoma, T.: Studien über bakteriophagenwirkung, Centralbl. f. Bakteriol., 1, O., 1924, 91, p. 554.

[200] Israelsky, W.: Bakteriophagie und Pflanzenkrebs, Centralbl. f. Bakteriol., 2, 1926, 67, p. 236.

[201] Jaumain, D.: Autolyse microbienne en tubes scellés, Compt. rend. Soc. de biol., 1922, 87, p. 790.

[201a] Jaumain, D., and Meullman, M.: Absorption du principe lytique par les microbes tues, Compt. rend. Soc. de biol., 1922, 87, p. 362.

[202] Joannides, G.: Sur l'action de la bile (ou le taurocholate de soude), sur le phénomène de la lysis en géneral, Compt. rend. Soc. de biol., 1924, 90, p. 40.

[203] Jonesco-Mihaiesti, C.: Studies on the Twort-d'Herelle Phenomenon, J. Exper. Med., 1924, 40, p. 317.

[204] Jötten, K.: Ueber das sogenännte d'Herellechen Phänomen, Klin. Wchnschr., 1922, 1, p. 2181.

[205] Kabeshima, T.: Sur un ferment d'immunité bactériolysant, du mecanisme d'immunité infectieuse intestinal; de la nature du dit "microbe filtrant bactériophage" de d'Hérelle, Compt. rend. Soc. de biol., 1920, 83, p. 219.

[206] Kabeshima, T.: Sur le ferment immunité bactériolysant, Compt. rend. Soc. de biol., 1920, 83, p. 471.

[207] Kasarnowsky, Sophie: Zur Frage des d'Hérelle-Phänomens, Ztschr. f. Hyg., 1926, 105, p. 504.

[208] Kasarnowsky, Sophie; and Timokin-Schuekoff, R.: Ueber die antigen Eigenschaften des bakteriophagen Lysins, Ztschr. f. Hyg., 1925, 104, p. 119.

[209] Katzu, S.: Bakteriophagähnliche Erscheinungen bei Milzbrand, Centralbl. f. Bakteriol., 1925, 96, p. 281.

[210] Kauffman, F.: Ueber die Beziehungen zwischen dem d'Herelleschen Lysin und den Autotoxinen (Conradi-Kurpjuweit), Ztschr. f. Hyg., 1926, 106, p. 308.

[211] Kauffman, F.: Keimwandlung und Lysinwirkung, Ztschr. f. Hyg., 1926, 106, p. 520.

[212] Keller, W.: Ueber Lysin und Trypsin (Ein Beitrag zur Biologie des Twort-d'Herelle Phänomens), Ztschr. f. Hyg., 1924, 103, p. 177.

[213] Kimura, S.: Ueber Schleimbildung bei Bakterien unter dem Einflusse von Bakteriophagen, Ztschr. f. Immunitätsforsch. u. exper. Therap., 1925, 42, p. 507.

[214] Koch, K.: Untersuchungen über Bakteriophagenkataphores, Centralbl. f. Bakteriol., 1926, 99, p. 209.

[215] Koch-Ziegenspock: Die Pettenkoferein als Erzeuger des d'Herelleschen Phänomens, Centralbl. f. Bakteriol., 1927, 71, p. 433.

[215a] Koch, M.: Die Kuhnschen Bakteriophagen, Bot. Arch., 1927, 19, p. 275.

[216] Koser, S. A.: Action of the bacteriophage on a thermophilic bacillus, Proc. Soc. Exper. Biol. & Med., 1926, 24, p. 109.

[217] Koser, S. A.: Transmissible lysis of a thermophilic organism, J. Infect. Dis., 41, p. 365.

[218] Kramer, S. P.: Bacterial filters, J. Infect. Dis., 1927, 40, p. 343.

[219] Krestownikowa, W., and Gubin, W.: Die Verteilung and die Ausscheidung von Bakteriophagen im Meerschweinchen-organismus bei subkutaner Applicationsart, J. Microbiol., Patolog. i. Infekzionnich bolesney, 1925, 1, p. 3; referred to in Centralbl. f. Bakteriol., 1, ref., 1926, 82, p. 283.

[220] Kropveld, S.: Studies over den Bacteriophaag tegen staphylokokken, Nederl. Tijdschr. v. Geneesk., 1923, 67, p. 1228; cited from d'Herelle.[193]

[221] Kuhn, P.: Weitere Einblick in die Entwicklung der A-Formen (Pettenkoferiaformen) Centralbl. f. Bakteriol., 1924, 93, p. 280.

[222] Kuhn, P.: Demonstration der Ergebnisse morphologische Bakterienstudien und zum d'Herelleschen Phanomen, Arch. f. Schiffs- u. Tropen- Hyg., 1926, 30, p. 133.

[223] Kuhn and Ebeling: Ztschr. f. Immunitätsforsch. u. exper. Therap., 1916, 25, p. 1; cited from Otto and Sukennikowa.[271]

[223a] Kuhn, Gildemeister and Woithe: On paragglutination, Arb. a. d. k. Gsndhtsamte, 1911, 31, p. 38.

[224] Kuhn and Woithe: On paragglutination, Med. Klin., 1909, p. 1709.

[225] Kuttner, Anne: Bacteriophage phenomena, J. Bact., 1923, 8, p. 49.

[226] Lacassagne, A., and Paulin, A.: Destruction du principe bactériolytique par les rayonnements corpusculaires du rayon, Compt. rend. Soc. de biol., 1925, 93, p. 1502.

[227] Larkum, N. W.: Bacteriophagy in urinary infections: I. The incidence of Bacteriophage and of Bacillus coli susceptible to dissolution by the bacteriophage in urines; presentation of cases of renal infection in which bacteriophage was used therapeutically, J. Bact., 1926, 12, p. 203.

[228] Larkum, N. W.: Bacteriophagy in urinary infections: II. Bacteriophagy in the bladder, J. Bact., 1926, 12, p. 225.

[229] Lesbre, P.: Bactériophage et anatoxine dans la vaccination antidysentérique par voie buccale, Compt. rend. Soc. de biol., 1925, 93, p. 614.

[230] Lepper, E. H.: The reproduction of bacteriophage when the sensitive organism is grown in a synthetic medium, Brit. J. Exper. Pathol., 1924, 5, p. 40.

[231] Levaditi, C., and Nicolau, S.: Propriétés physiques des ultravirus neurotropes, Compt. rend. Soc. de biol., 1923, 88, p. 66.

[232] Levy, M.: Essai de traitement de la typhose murine par le bactériophage, Compt. rend. Soc. de biol., 1925, 93, pp. 82, 395.

[233] Lisbonne, M.; Boulet and Carrère: Sur l'obtention du principe bactériophagique au moyen d'exudates leucocytaires in vitro, Compt. rend. Soc. de biol., 1922, 86, p. 340.

[234] Lisbonne, M., and Carrère, L.: Sur l'obtention du principe bactériophagique par antagonisme microbienne, Compt. rend. Soc. de biol., 1922, 87, p. 1011.

[235] Lisbonne, M., and Carrère, L.: Influence des électrolytes sur la lyse microbien transmissible, Compt. rend. Soc. de biol., 1923, 89, p. 865.

[236] Lisbonne, M., and Carrère, L.: Sur apparation spontanée du pouvoir lysogène dans les cultures pures, Compt. rend. Soc. de biol., 1924, 90, p. 265.

[237] Louet, G.: The bacteriophage as an agent of vaccination against "barbone" disease, J. Am. Vet. M. A., 1925, 67, p. 713.

[238] Lucchini, C., and Villa, L.: Batteriofargo e tossine batteriche: natura della terapia col batteriofargo, Boll. d. Inst. sieroterap., 1926, 5, p. 231.

[239] Machardo, A., and da Costa Cruz, J.: Soble a autolyse microbiana transmissivel; batteriophargo de d'Herelle, Brasil-med., 1921, 35, p. 347; cited from d'Herelle.

[240] Maisin, J.: Au sujet de la nature du principe bactériophage, Compt. rend. Soc. de biol., 1921, 84, p. 467.

[241] Manniger, R.: Beitrag zur Kenntnis der Bakteriophagie, Centralbl. f. Bakteriol., 1926, 99, p. 203.

[242] Manoliu, E., and Costin: Eaux polluteés et lyse transmissible, Compt. rend. Soc. de biol., 1925, 93, p. 384.

[243] Marcuse, K.: Grundlagen und Aufgaben der Lysintherapie (d'Herelle's Bakteriophagen), Deutsche med. Wchnschr., 1924, 50, p. 334; cited from Larkum.

[244] Marcuse, K.: Untersuchungen über das d'Herelellesche Phänomen: II. Untersuchungen über die Bedeutung der Leukocyten für das d'Herelleschen Phänomen., Ztschr. f. Hyg., 1924, 102, p. 206.

[245] Marcuse, K.: Relation between bacteriophagic and serologic action, Ztschr. f. Hyg., 1926, 105, p. 17.

[246] Marshall, M.: Observations on d'Herelle's bacteriophage, J. Infect. Dis., 1925, 37, p. 126.

[247] Maslakowetz, P., and Kasarnowsky, Sophie: Versuche zur Herstellung von Antigenen mittel bakteriophagen Lysins, Mikrobiol. J., 1926, 3, pt. 2; referred to in Centralbl. f. Bakteriol., 1927, 86, p. 478.

[248] McKinley, E. B.: The relation of digestive enzymes and ferments to the phenomenon of d'Herelle, J. Bact., 1923, 8, p. 543.

[249] McKinley, E. B.: Further notes on d'Herelle's phenomenon, J. Lab. & Clin. Med., 1923, 9, p. 3.

[250] McKinley, E. B.: The bacteriophage in the treatment of infections, Arch. Int. Med., 1923, 32, p. 899.

[251] McKinley, E. B.: Serum antilytique obtenu par immunization contre un bacterie normale, Compt. rend. Soc. de biol., 1925, 93, p. 1050.

[252] McKinley, E. B.: Transformation, sous l'influence du principe lytique faible, de la spécificite antigenique d'une culture, Compt. rend. Soc. de biol., 1925, 93, p. 1052.

[253] McKinley, E. B.; Fischer, R., and Holden, M.: Action of ultraviolet light upon bacteriophage and filtrable viruses, Proc. Soc. Exper. Biol. & Med., 1926, 23, p. 408.

[254] McKinley, E. B., and Spense, R.: The therapeutic value of bacteriophage in the treatment of bacillary dysentery, South. M. J., 1924, 17, p. 563.

[255] Meissner, G.: Ueber Bakteriophagen gegen Choleravibrionen, Centralbl. f. Bakteriol., 1924, 91, p. 149.

[256] Meissner, G.: Der Bindungsverhältnisse zwischen Bakteriophage und Bakterien, Centralbl. f. Bakteriol., 1924, 93, p. 489.

[256a] Mellon, R. R.: The infectivity and virulence of a filtrable stage in the life history of B. fusiformis and related organisms, J. Bact., 1926, 12, p. 279.

[257] Montiro, L.: Présence du bactériophage dans l'eau des riviéres de San Paulo, Compt. rend. Soc. de biol., 1926, 95, p. 394.

[258] Nankamura, O.: Das Verhalten von Bakteriophagen in Gelatine, Wien. klin. Wchnschr., 1923, 36, p. 86.

[260] Nakashima, T.: Beitrag zur Vorkommen und Verhalten des Bakteriophagen Lysins im Abwassern, Centralbl. f. Bakteriol., 1, O., 1925, 94, p. 303.

[261] Nicolle, C.: Sur la nature des virus invisible; origin microbienne des inframicrobes, Arch. d. Inst. Pasteur de l'Afrique du Nord, 1925, 14, p. 105.

[262] Nobechi, K.: Sur la préparation du bactériophage pour le vibrion cholérique et la classification de ces vibrions au point de vue du phénomène de bactériophage, Compt. rend. Soc. de biol., 1926, 95, pp. 1250, 1252.

[263] Ogata, N.: Ueber die Beeinflussung biologisch-chemischer Eigenschaften der Bakterien durch Bakteriophagen, Ztschr. f. Immunitätsforsch. u. exper.. Therap., 1925, 45, p. 465.

[264] Okamoto, T.: Untersuchungen über die Beziehungen des d'Herelleschen Phänomens zum N-Stoffwechsel der Bakterien, Ztschr. f. Immunitätsforsch. u. exper. Therap., 1925, 42, p. 16.

[265] Olsen, O., and Strauss, W.: Untersuchungen zur Frage der Tropfenverstreuung mittels des d'Herelleschen Lysats. Ztschr. f. Hyg., 1926, 105, p. 552.

[266] Ordelt, V.: Der Einfluss der Reaktion auf das Bakteriophagum, intestinale und andere Versuche, Centralbl. f. Bakteriol., 1, O., 1924, 78, p. 519.

[267] Ørskov, J., and Larsen, A.: On bacterial variation, J. Bact., 1925, 10, p. 473.

[268] Osumi, S.: Serologischen Studien mit einen Bakteriophagen, Ztschr. f. Immunitätsforsch. u. exper. Therap., 1924, 40, p. 261.

[269] Otto, R., and Munter: Zum d'Herelleschen Phänomen, Deutsche med. Wchnschr., 1921, 47, pp. 1579, 52.

[270] Otto, R.; Munter and Winkler: Beiträge zum d'Herelleschen Phänomen, Ztschr. f. Hyg. 1922, 96, p. 118.

[271] Otto, R., and Sukennikowa: Bakteriophages Lysin und Paragglutination, Ztschr. f. Hyg., 1923, 101, p. 119.

[272] Otto, R., and Winkler: Ueber die Natur des d'Herelleschen Bakteriophagen, Deutsche med. Wchnschr., 1922, 48, p. 383.

[273] Pelouse, P., and Schofield, F.: The gonophage, J. Urol., 1927, 17, p. 407.

[274] Peieira, O.: O phenomeno de d'Herelle e as infeccoes pelos bacilos coli e dysentericos, 1924; cited from d'Herelle.

[275] Pesch, K.: Milzbrand Pseudo-bakteriophagen, Centralbl. f. Bakteriol., 1, O., 1925, 93, p. 525.

[276] Petrovanu, G.: Recherches sur la présence du principe lytique vis-a-vis du vibrion cholérique dans la paroi de la intestin grêle, Compt. rend. Soc. de Biol., 1924, 91, p. 754.

[708] Petrovanu, G.: Sur la présence du principe lytique dans l'éxudat amygdalien de diversés angines, Compt. rend. Soc. de biol., 91, p. 502.

[277] Petrovanu, G.: Récherches sur l'existence du principe lytique dans la péritonite cholérique expérimentale, Compt. rend. Soc. de biol., 1924, 91, p. 735.

[278] Philibert, A.: Le principe bactériophage, Médecine, 1923, 5, p. 91.

[279] Pico, C.: Autolyse transmissible sans l'intervention de l'hypothétique virus bactériophage, Compt. rend Soc. de biol., 1922, 87, p. 836.

[280] Pico, C.: Le principe lytique, est-il contenu dans les bactéries? Compt. rend. Soc. de biol., 1922, 87, p. 687.

[281] Pico, C.: Autolyse transmissible du B. anthracis sans intervention de l'hypothétique virus bactériophage, Compt. rend. Soc. de biol., 1922, 87, p. 836.

[282] Pierrot, R., and Bilouet, V.: Le bactériophage de d'Hérelle chez le nouveau-ne, Compt. rend. Soc. de biol., 1925, 93, p. 635; 1926, 95, p. 301.

[283] Pondman, A.: Proeven tot het verkrijgen van het verschijnsel van d'Herelle uit rein-culturen, 1923; cited after d'Herelle.

[284] Prausnitz, C.: Ueber die Natur des d'Herelleschen Phänomens, Klin. Wchnschr., 1922, 1, p. 1639.

[285] Prausnitz, C.: Untersuchungen über den d'Herelleschen Bakteriophagen: I. Die Natur des Bakteriophage, Centralbl. f. Bakteriol., 1, O., 1922, 89, p. 187.

[286] Prausnitz, C., and Firle, E.: Neue Untersuchungen über das Wesen des Bakteriophagen, Centralbl. f. Bakteriol., 1, O., 1924, 93, p. 148.

[287] Prausnitz, C., and van der Reis, V.: Untersuchungen des menschlichen Dunndarmin-haltes auf Bakteriophagen, Deutsche med. Wchnschr., 1925, 51, p. 304.

[288] Von Preisz, H.: Die Bakteriophage, vornehmlich auf grund eigener Untersuchungen, 1925.

[289] Proca, G.: Sur la résistance du principe lytique (bactériophage) au bichloure de mer-cure, Compt. rend. Soc. de biol., 1927, 96, p. 1244.

[290] Proca, G.: La bactériolyse d'origine amebienne et le phénomène de d'Hérelle, Compt. rend. Soc. de biol., 1926, 95, p. 143.

[291] Pyle, N.: The bacteriophage in relation to Salmonella pullora, J. Bact., 1926, 12, p. 245.

[292] Radziewsky, A.: Untersuchungen zur Theorie der bakteriellen Infektion, Ztschr. f. Hyg., 1901, 37, p. 1.

[293] Reynals, F.: Bactériophage et microbes tués, Compt. rend. Soc. de biol., 1926, 94, p. 242.

[294] Richet, C.; Aserad and Delarve: Sur un cas de fièvre typhoïde ataxo-adynamique trés rapidement guéri par un stock-bactériophage anti-Éberth., Bull. et mém. Soc. méd. d. hôp. de Paris, 1924, 40, p. 838.

[295] Richet, C., and Hauduroy: Essais d'immunizations contre la typhose des souris avec le bactériophage de d'Hérelle, Compt. rend. Soc. de biol., 1925, 93, p. 222.

[296] Rivers, T.: The effect of repeated freezing (—185 C.) and thawing on colon bacilli, virus III, vaccine virus, herpes virus, bacteriophage, complement and trypsin, J. Exper. Med., 1927, 45, p. 11.

[297] Rosenthal, L.: Sur lysobactéries thérmophiles, Compt. rend. Soc. de biol., 1925, 93, p. 1569.

[298] Sanderson, E.: A note on the bacteriophage with reference to Complement-Fixation Tests, J. Immunol., 1925, 10, p. 625.

[299] Sangiorgi, G., and Vercellana, G.: Il "principio lytico" nelle acque di alcuni fiumi italiani, Ann. d'ig., 1926, 84, p. 425.

[300] Schnabel, A.: Uebertragung allergischen Zustände bei Bakterien; Ein neuer Gesichts-punkt fur das Twort-d'Herelle Phänomen, Klin. Wchnschr., 1924, 37, p. 566.

[301] Schuurman, C.: Der Bakteriophage, ein lebender Organismus, Centralbl. f. Bakteriol., 1925, 95, p. 97.

[302] Schutze, H.: The permanence of the serological paratyphoid B types with observations on the non-specificity of agglutination with "rough" variants, Brit. J. Hyg., 1922, 30, p. 330.

[303] Schwartzmann, G.: Observations on lytic principle of weak potency, Proc. Soc. Exper. Biol. & Med., 1925, 22, p. 433.

[204] Schwartzmann, G.: The influence of partial anaerobiosis upon regeneration of a highly diluted lytic principle, J. Exper. Med., 1925, 42, p. 507.

[305] Schwartzmann, G.: The influence of oxygen on the behavior of Bacillus coli towards lytic principle, J. Exper. Med., 1926, 43, p. 743.

[306] Schwartzmann, G.: La réduction du blue de méthylene dans l'autolyse transmissible, Compt. rend. Soc. de biol., 1926, 95, p. 431.

[307] Schwartzmann, G.: The rate of reduction of methylene blue by Bacillus coli in the course of the bacteriophage phenomenon, Centralbl. f. Bakteriol., 1926, 101, p. 62.

[308] Schwartzmann, G.: Studies on the streptococcus bacteriophage: I. A powerful lytic principle against hemolytic streptococci of erysipelas origin, J. Exper. Med., 1927, 46, p. 497.

[309] Seiffert, W.: Neue Untersuchungen über den Charakter des d'Herelleschen Phänomens, Med. Klin., 1923, 19, p. 833; Ztschr. f. Immunitätsforsch. u. exper. Therap., 1923, 38, p. 292.

[310] Seiffert, W.: Das d'Herellesche Phänomen als "exogene Autolyse der Bakterien," Ztschr. f. Hyg., 1922, 98, p. 482.

[310a] Seiffert, W.: Zum d'Herelleschen Phänomen, Klin. Wchnschr., 1923, 2, pp. 1049 and 1479.

[311] Seiffert, W.: Ueber das d'Herelleschen Phänomen, Soziale Hyg. u. Mikrobiol., Sept. 21, 1922; cited from d'Herelle.

[312] Sierakowsky, S., and Zajdel, R.: Sur les baktéries provoquant la carie dentaire, Compt. rend. Soc. de biol., 1924, 90, p. 961.

[313] Smith, J.: The bacteriophage in the treatment of typhoid fever, Brit. M. J., 1924, 47, p. ???.

[314] Sonnenschein, C.: Zur Kenntniss bakteriophagähnlichen Erscheinungen, Centralbl. f. Bakteriol., 1925, 95, p. 257.

[315] Sonnenschein, K.: Die Herkunft der Bakteriophagen im Menschen und im tierischen Organismus, Gior. di. batteriol. e immunol., 1927, 2, p. 1.

[315a] Sonnenschein, C.: Neue Gesichtspunkte der Bakteriologie und Ätiologie der Rhinitis atrophicans chronica (foetida), Folia Oto-Laryngol., 1925, p. 450.

[316] Stassano, H., and Beaufort, A.-C.: Le principe lytique transmissible soumis au criterium de ultrafiltration ou filtration moléculaire, Compt. rend. Soc. de biol., 1925, 93, p. 1378.

[317] Stassano, H., and Beaufort, A.-C.: Action de citrate du soude sur le principe lytique transmissible, Compt. rend. Soc. de biol., 1925, 93, p. 1380.

[318] Stassano, H., and Beaufort, A.-C.: L'action de l'ether sur le principe lytique transmissible, Compt. rend Soc. de biol., 1925, 93, p. 1382.

[319] Stolz, J.: Agglutination and the bacteriophage in typhoid, Čas. lék. česk., 1925, 64, 64, p. 1588.

[320] Suramyi and Kramar: Ueber das Vorkommen des d'Herelleschen Bakteriophagen in Sauglingsstulen, Monatschr. f. Kinderh., 1923, 26, p. 392.

→ [321] Topley, W.; Wilson and Lewis: The rôle of the Twort- d'Herelle phenomenon in epidemics of mouse typhoid, J. Hyg., 1925, 24, pp. 17 and 295.

[322] Twort, F.: An investigation into the nature of ultramicroscopic viruses, Lancet, 1915, 2, p. 124.

[323] Urech, E., and Pache, H.: Contribution a l'étude des sérums anti-bactériophages, Schweiz. med. Wchnschr., 1926, 56, p. 275.

[324] Van Loghem, J.: Bakteriophage und endolytisches endotoxin des Choleravibrio, Centralbl. f. Bakteriol., 1, O., 1926, 100, p. 19.

[325] Vignati, J.: Antagonisme entre le bacille typhique et le B. coli, Compt. rend. Soc. de biol., 1926, 94, p. 212.

[326] Villazon, N.: Bactériophage efficacé contre le bacille de la peste, Compt. Soc. de biol., 1923, 89, p. 754.

[327] Wagemans: Bull. Acad. roy de méd. de Belg., 1923; cited from Wollman.[328]

[328] Watanabe, T.: Serologischen Untersuchungen an Shigabakteriophagen, Ztschr. f. Immunitätsforsch. u. exper. Therap., 1923, 37, p. 106.

[328a] Watanabe, T.: Desinfectionsversuche mit Bakteriophagen, Arch. f. Hyg., 1923, 92, p. 1.

[329] Weinberg, M., and Aznar, P.: Autobacteriolysines et le phénomene de d'Herelle. Compt. rend. Soc. de biol., 1922, 86, p. 833.

[330] Weiss, E.: The bacteriophage-antibacteriophage reaction, J. Immunol., 1927, 13, p. 301.

[331] Weiss, E.: Bacteriophage purified with lipoids, J. Immunol., 1927, 13, p. 311.

[332] Wolff, L.: Ztschr. f. Immunitätsforsch. u. exper. Therap., 1926, 45, p. 223.

[332a] Wolff, L., and Jansen, J.: Action de divers antiseptiques sur le bactériophage de d'Herelle, Compt. rend. Soc. de biol., 1922, 87, p. 1807.

[333] Wollman, E.: Sur le phénomène de d'Herelle, Compt. rend. Soc. de biol., 1921, 84, p. 3.

[334] Wollman, E.: Recherches sur le phénomène de d'Hérelle; action de la trypsin sur le bactériophage du bacille Shiga, Compt. rend. Soc. de biol., 1924, 90, p. 59.

[335] Wollman, E.: Récherches sur le Bactériophagie (phénomène de Twort-d'Hérelle), Ann. de l'Inst. Pasteur, 1925, 39, p. 789.

[336] Wollman, E.: Récherches sur la bactériophage (phénomène de Twort-d'Hérelle), Ann. de l'Inst. Pasteur, 1927, 41, p. 883.

[337] Wollman, E., and Brutsaert, P.: L'autonomy des bactériophages, Compt. rend. Soc. de biol., 1925, 92, p. 1284.

[338] Wollman, E., and Goldenberg, L.: Le phénomène de d'Hérelle et la réaction de fixation, Compt. rend. Soc. de biol., 1921, 85, p. 772.

[239] Wollman, E., and Reynals, F.: Bactériophage et autolyse, Compt. rend Soc. de biol., 1926, 94, p. 1330.

[240] Wollman, E., and Suarez, E.: Ultrafiltration du bactériophage et des proteines sériques, Compt. rend. Soc. de biol., 1927, 96, p. 15.

[241] Wollman, E., and Wollman, E.: Récherches sur la phénomène de d'Hérelle; pluralité et autonomie des bactériophages, Compt. rend. Soc. de biol., 1925, 92, p. 552.

[242] Wollman, E., and Wollman, E.: Sur la transmission "para-héréditaire" de charactéres chez les bactéres, Compt. rend. Soc. de biol., 1926, 93, p. 1568.

[243] Wollman, E., and Wollman, E.: Fixité et variabilité des characteres chez les bactériophages; properties antigenic, Compt. rend. Soc. de biol., 1927, 96, p. 332.

[244] Wollstein, M.: Studies on the bacteriophage of d'Herelle with Bacillus dysenteriae, J. Exper. Med., 1921, 34, p. 467.

[245] Zdansky, E.: Gewinnung specifischer Bakteriophagen und über bacteriophagentherapeutische Versuche, Med. Klin., 1924, 20, p. 1485.

[246] Zdansky, E.: Ueber die Bedeutung der Salze für die übertragbaren Lysine (Bakteriophagen), Wien. klin. Wchnschr., 1924, 37, p. 141.

[247] Zinsser, H., and Tyzzer, E. E.: Textbook of Bacteriology, 1927.

[248] Zinsser, H., and Tang, Fei-Fang: Studies on Ultrafiltration, J. Exper. Med., 1927, 46, p. 357.

[249] Zoeller, C., and Manoussakis: Kerato-conjunctivité experimentale a bacille pyocyanique et l'action d'un bactériophage anti-pyocyanique, Compt. rend. Soc. de biol., 1924, 91, p. 548.

[249a] Zoeller, C., and Manoussakis: Étude sur le bactériophage en sacs de collodion, Compt. rend. Soc. de biol., 1925, 93, p. 1091.

[250] Zoeller, C.: Action des rayons ultraviolets sur une souche de bactériophage, Compt. rend. Soc. de biol., 1923, 89, p. 860.

[251] Personal communication to the author.

[252] For complete reference see bibliography in Microbic Dissociation, Hadley.[145]

ADDENDA

[353] Bazy, L.: Traitement des infectiones chiurgicales à staphylocoques par le bactériophage anti-staphylocoque, Compt. rend. Soc. de biol., 1925, 92, p. 485.

[354] Bechold: Ztschr. f. Hyg., 1926, 105, p. 601; 106, p. 580.

[354a] Bronfenbrenner, J.: The study of the intimate mechanism of the lysis of bacteria by bacteriophage, Am. J. Pathol., 1927, 3, p. 562 (abstract).

[354b] Bronfenbrenner, J.: The particulate nature of the bacteriophage, Am. J. Pathol., 1927, 3, p. 561 (abstract).

[354c] Bronfenbrenner, J.: The particulate nature of the bacteriophage, J. Exp. Med., 1927, 45, 873.

[355] Broudin, L.: Note préliminaire sur le bactériophage de la Pasteurella aviare, Compt. rend. Soc. de biol., 1926, 95, p. 1332.

[356] Fejgin, B.: Sur la nature de phénomène de d'Hérelle, Compt. rend. Soc. de biol., 1923, 89, p. 1385.

[357] Fränkel, E., and Schultz: Beiträge zur Frage des Bakteriophagen (d'Herellesches Lysin), Ztschr. f. Immunitätsforsch. u. exper. Therap., 1927, 51, p. 382.

[358] Grasset, E.: Récherches sur le passage du bactériophage à travers le placenta, Compt. rend. Soc. de biol., 1927, 96, p. 839.

[359] Grumbach, A., and Dimtza, A.: Die Bedeutung des Bakteriophagen für die bakteriologische Diagnostik, Ztschr. f. Immuntätsforsch. u. exper. Therap., 1927, 51, p. 176.

[360] Von Jeney, A.: Bacteriophage Erscheinungen an einem Wasserbakterium, B. cloacae, Centralbl. f. Bakteriol., 1927, 102, p. 263.

[361] Kline, G. M.: The demonstration of bacteriophage in old stock cultures, J Am. Pub. Health A., 1927, 12, p. 1074.

[362] De Neckar, J.: De l'absorption du principe bactériophage par les colloïdes, Compt. rend. Soc. de biol., 1922, 87, p. 1247.

[363] Nungester, W.: Dissociation in B. anthracis, Proc. Soc. Exper. Biol. & Med., 1927, 24, p. 959.

[364] Otto, R., and Munter: Weichardt's Jahresb., 1924, 6, p. 123.

[365] Pesch, K., and Sonnenschein, C.: Variabilität und Bakteriophagen bei Pyocyaneusbakterien, Klin. Wchnschr., 1925, 4, p. 1585.

[366] De Poorter, P., and Maisin, J.: Contribution a l'étude de la nature du principe bactériophage, Arch. internat. de pharmacol., 1921, 25, p. 473; cited from d'Herelle.[196]

[367] Prausnitz, C.: Die Bakteriophagen, Handbuch d. Exper. Therap., 1926.

[368] Schumacher: Ueber den Nachweis des Bakterienkerns und Seine chemische Zusammensetzung, Centralbl. f. Bakteriol., 1, O., 1926, 97, p. 81. Also: J. Chem. d. Zelle u. Gewebe, 1926, 1, p. 1.

[369] Sonnenschein, C.: Ueber Paratyphus-Bakteriophagen und Antiphagine, Centralbl. f. Bakteriol., 1926, 97, p. 312.

[370] Sonnenschein, C.: Atypische Wuchsformen von Bakterien als Krankheitserreger. Mucosus-Form von Paratyphi im Blut eines Paratyphuskranken, Centralbl. f. Bakteriol., 1926, 100, p. 11.

[371] Wedemann: Centralbl. f. Bakteriol., 1926, 97, p. 50.

[372] Weiss, E.: The behavior of the bacteriophage in sugar media, J. Lab. & Clin. Med., 1927, 12, p. 937.